FIRST 100 YEARS
1902-2002

Stewart & Stevenson
A Look Back At The First 100 Years
1902 - 2002

by
Myrtle D. Malone

published by
Pioneer Publications, Inc.
Bill Levisohn, Publisher

Pioneer Publications, Inc.
30652 Via Estoril
Laguna Niguel, California 92677

ISBN: 1-881547-30-2
Library of Congress Number: 2001116127
Pioneer Publications, Inc., Houston, Texas 77070

Photographs courtesy of Stewart & Stevenson Services, Inc.

Contents

SIDEBAR STORIES

ILLUSTRATIONS

Acknowledgments

The author wishes to thank the following people associated with Stewart & Stevenson for their help in making this book possible:

Mr. C. Jim Stewart III, Vice President of Public Affairs and Marketing, who provided the critical leadership, review and direction, and unlimited access to company archives

Mr. Dan Kaiser, who pulled together vast amounts of research material and photographs, and was my day-to-day contact

Mr. C. Jim Stewart II, Chairman of the Board, whose open-door policy allowed me ample interview time to delve into family and company history

Ms. Madlin Stevenson, who graciously provided the history of the Stevenson family and memories of "horse and carriage" Houston in the early part of the century, and who, sadly, passed away during the writing of this history

Mr. Joe Manning, retired President, who tirelessly provided the history of his 47 years with the company, reviewed my copy, and shared pictures and other items from his personal collection

Mr. Donald Stevenson, Mr. Keith Stevenson, Mr. Mike Andrews, Mr. Garth Bates, Mr. C. Jim Stewart III, Mr. Rick Stewart, and Mr. David Stewart, who provided memories of their "growing up" years with the business, as well as insight into various operations of the company which they personally influenced in positions of leadership

Mr. C. LaRoy Hammer, retired Vice President of Government Contracts, who provided the background story of how the FMTV program came about

Mr. Dick Connolly, retired Vice President of Branches and Dealers Worldwide; Mr. E.E. "Slim" Childress, retired Vice President of Service; Linda Langham Cole, Susan Langham Mendenhall and Meridith Mendenhall, wife, daughter and granddaughter of Tom Langham; and John C. Metcalf, retired, who provided additional sidebar stories about people and events

Patsy Andrews, wife of Mike Andrews, whose research of early Houston directories provided additional information about the Stewart and Stevenson families, as well as C. Jim Stewart & Stevenson

Doyleen Harris, Executive Assistant to Mr. C. Jim Stewart II, who kindly set up interview appointments, provided typed copies of history dictated by Joe Manning, and more than once provided information that solved some confusion on my part

My personal thanks also to my husband, Max, and to our sons, Geoffrey, Philip and Kevin, for their support and encouragement, especially their tolerance of my many Saturdays and late evenings in the office during the course of researching, writing and designing this book.

Foreward

The story of Stewart & Stevenson is the story of two families who joined forces and founded a company on a set of basic and enduring principles that have enabled it to progress through 100 years and achieve the American Dream.

C. Jim Stewart & Stevenson, the $600 blacksmith and carriagemaker "pardnership" that catered to the horse and carriage trade a century ago today is a billion-dollar corporation that builds the U.S. Army Family of Medium Tactical Vehicles utilized in the nation's defense.

Led by generations of visionary Stewarts and Stevensons, the company progressed from horseshoes to horsepower, Diesel power, and gas turbine power, and aided by the ingenuity of a family of dedicated employees, the company created engineered power applications for the agriculture, marine, aviation, power generation, defense, transportation and petroleum industries. Behind each product developed, over the years and still today, stands the company's "anytime, anywhere" parts and service total responsibility.

Guiding Stewart & Stevenson into its second century is new management with new goals. Still holding true to the founders' basic and enduring principles, Stewart & Stevenson and its family of 4,200 employees look toward the future with power and vision.

Prologue

In 1902 my grandfather, C. Jim Stewart, a master blacksmith, owned a horseshoeing parlour here in Houston. He envisioned a better future for the business if he were able to supply more of the needs of Houston's bourgeoning horse and carriage trade. With that goal in mind, he formed a "pardnership" with Joe R. Stevenson, the "carriagemaker," and the firm of C. Jim Stewart & Stevenson was born in April 1903.

Each of the "pardners" contributed $300, and a partnership agreement was drawn up. It specified my grandfather's duties as the blacksmith, Mr. Stevenson's duties as the woodworker, and included the proviso that "both shall do such things in and about said business which shall be necessary." Concluded with signatures and a handshake, the agreement set the standard for teamwork, integrity, and a dogged determination to get the job done right — the basic principles which have guided our company through 100 years of our history, and to which we still adhere today.

C. Jim Stewart & Stevenson originated as a family business, and from the beginning the employees were considered an integral part of the family. Every employee was expected to work five and a half days a week, and were paid at noon on Saturday. Although employee benefits were unheard of in those days, it was unwritten company policy to help employees and their family members who became ill, and those who became too sick or too old to work were taken care of as part of the family. For many years, our management knew every employee's family members by name. While our astounding growth has long since made that a lot more difficult, the sense of family still remains. Policies initiated over the years, such as tuition reimbursement that enabled many employees to achieve a higher level of education and the summer training program for children of employees, have contributed to the sense of family. The company picnic, which originated in the horse and buggy days, is still an annual event attended by our employees and their families.

Through hard work, integrity, and visions of better products, better service, and better ways to conduct business, successive generations of the Stewart & Stevenson's "family" of employees have made this an enduring and prosperous company. Like ambassadors of good will, they have also made friends of our business associates, both suppliers and customers alike, who have greatly contributed to the success and growth of our business.

The $600 family "pardnership" begun by C. Jim Stewart and Joe R. Stevenson is today a global, billion-dollar, public corporation. To grow from such humble beginnings into an ongoing, successful business — still following our founders' basic principles 100 years later — is indeed something of which all of us can be very proud.

I began my career with the company working part time in the summers of my youth, then full time in 1946. I've seen incredible changes take place in our organization in the last half century. I've seen several generations of our employees, along with my own family, which includes sons and grandsons now working with the company, grow with the company. As I retire at the end of our first century in business, I leave the company in great hands — leaders who will guide this company to even greater heights in its next century.

We celebrate our 100th anniversary with great joy — and with eager anticipation for an even greater future as our second century begins.

C. Jim Stewart II, Chairman

C. Jim Stewart II
Chairman of the Board
Grandson of Company Co-Founder

Turn of the Century 1890 - 1902

C. Jim Stewart & Stevenson
Est. 1902
Houston, Texas

Out of This Time and Place

As the last years of the final decade of the 1800s rolled toward a new century, Houston, Texas bustled with activity. Workers streamed into plants along the Buffalo Bayou where they produced beer and cigars, tiles and textiles, wagons and carriages. Market Square, the Congress-Travis-Preston-Milam block, teemed with farmers from outlying areas on buckboards loaded with produce and driving livestock for sale. Packing houses and lumber yards did a brisk trade. Cotton compresses handled thousands of bales through the city's port, where funds that trickled down from the nation's capitol gradually financed the deepening of the bayou's channel to the Gulf. The iron horses of the Southern Pacific Railroad, headquartered nearby, arrived daily with families, investors and adventurers. By 1900, the city's population was approaching 45,000.

Among those arriving was John Henry Kirby, founder of Kirby Lumber Company, who settled into a brick Victorian home located at Smith and Gray. His ambitions for Houston to become a major industrial center were launched when oil gushed forth at Spindletop on the first day of January, 1901. Another arrival was Jesse H. Jones, who soon owned a chain of lumber yards and began constructing buildings around town. His lumber and construction enterprises rapidly expanded into banking and newspapering.

Houston at the turn of the 20th century had a dozen schools and forty-four churches. Merchant William Marsh Rice, determined that the city should have an institution of higher education, chartered an institute in his name and endowed it with $200,000, although its opening would be delayed until 1912. Legal complications following the death of Rice's wife in 1896, and Rice's murder in 1900 at the hands of Charles A. Jones, his secretary, first had to be cleared. Rice's lawyer, James A. Baker, the son of the co-founder of the Baker and Botts law firm, exposed a conspiracy between Jones and Albert T. Patrick, a Houston lawyer who had fled to New York City to avoid disbarment. Patrick worked as an agent for the Houston attorney defending Mrs. Rice's will, which Rice had challenged, and had written a bogus will that purportedly would leave him the lion's share of Rice's estate. It was Patrick's and Jones' intentions, had they succeeded in their villainy, to split the proceeds.

Houston in 1900 also had its saloons and brothels. One Bob Smith was the proprietor of a string of knock-down, dry-out saloons, the worst said to begin on the site of the Old Katy Depot where barmen kept shaker-spout bottles of knockout drops on the back bar to aid them in rolling railroad men on payday. One of Smith's bartenders, known as Mustache Jim, was married to a madam called Big Annie. Big Annie housed 20 girls in a two-story, red brick building on the southeast corner of Congress and Chenevert. The building had 22 rooms — the extra two were occupied by Mustache Jim, Big Annie and their two children.

By the turn of the century, electric arc lights had replaced gaslights along the city's streets, and the Houston Gas Company was busy convincing Houston's housewives to switch to fashionable gas ranges for cooking. But the ultimate status symbol of the day was the carriage, many of which were manufactured locally. Houston's elite made great display of their fortunes through their finely appointed carriages. Thoroughbred horses clopped over the city's wooden block streets carrying the uppercrust about town, while Sunday horse and buggy riders populated the new oyster shell road to Harrisburg. By all accounts, it was an impressive parade.

From Horseshoes to Horsepower® 1902 - 1919

A New Enterprise Emerges

Out of this scenario emerged a horse-shoer named C. Jim Stewart who, in 1894, was working as a blacksmith at the Houston Carriage Company, corner of Milam and Commerce. Though a somewhat small man, C. Jim (the C. stood for Cornelius) was a "scrapper" who worked out with his friend Bob Fitzsimmons, then the world's middle-weight boxing champion. In return, Fitzsimmons, who was in Houston training for his fight with James J. Corbett, would help C. Jim shoe horses.

About this same time, a young man named Joseph R. Stevenson was serving a woodworker apprenticeship at the Mosehart & Keller Carriage and Wagon Factory, 101 San Jacinto. They were not acquainted at the time, but Joe had heard of C. Jim. A softspoken gentleman, Joe appeared to be the opposite of C. Jim, who reportedly took no guff off of mean mules or complaining draymen. There were some who thought C. Jim could have been a boxer himself.

In 1902, C. Jim opened his own blacksmith and carriage shop at 1718 Congress Avenue, just one door west of Big Annie's house of ill fame. Joe R. Stevenson, who had become a fine woodworking craftsman, came to work for C. Jim in the carriage shop. In 1903, C. Jim offered young Stevenson a half interest in the business. Joe accepted.

Each man invested $300 in the new firm, which was to be called C. Jim Stewart & Stevenson. Henry J. Dennenbaum drew up the partnership agreement dated April 4, 1903. It clearly defined each partner's respective duties:

"The said C. J. Stewart shall do the general blacksmith work connected with said business, and said Joe R. Stevenson shall do the woodwork, but both shall do such things in and about said business, which shall be necessary."

From that day forward, Joe referred to himself as the "pardner" and signed that "title" after his name. In addition to his carriage work, Joe took charge of the books. They agreed that neither partner would draw more than $15 per week out of the firm's funds.

THE STATE OF TEXAS,

COUNTY OF HARRIS. KNOW ALL MEN BY THESE PRESENTS:--

That I, C. J. Stewart, of Harris County, for and in consideration of the sum of three hundred($300.00) dollars to me paid and to be paid by Joe R Stevenson, as follows: One hundred and fifty($150.00) dollars cash, the receipt of which is hereby acknowledged, and the balance of one hundred and fifty($150.00) dollars to be payable on or before twelve months after this date, without interest, have this day bargained, sold and conveyed, and by these presents do bargain, sell and convey unto the said Joe R Stevenson an undivided one-half(1/2) interest in and to all the tools now contained in the blacksmith and carriage shop conducted by me at 1712 Congress Avenue; also all the stock and material in said buildings; also the buildings on said premises; also the lease of said premises now held by me, including each and every article and character of property contained in and about said premises, and I hereby warrant that I am the owner of said property and that the same is unincumbered, and I further agree to pay all debts now due or owing by me and incurred in conducting the aforesaid business. I also hereby convey one-half the good will of the business heretofore conducted by me at said place.

Witness my hand at Houston, Texas, this April 4th, 1903.

C. J. Stewart

The State of Texas,

County of Harris. Before me, Henry J Dannenbaum, a notary public in and for Harris County Texas, on this day personally appeared C. J. Stewart, known to me to be the person whose name is subscribed to the foregoing instrument of writing and acknowledged to me that he executed the same for the purposes and considerations therein expressed.

Given under my hand and seal of office this April 4th, 1903.

Henry J. Dannenbaum

Notary Public, Harris County, Texas.

C. Jim Stewart II Recalls C. Jim I

"My grandfather thought the world of Joe Stevenson," C. Jim Stewart II later recalled. "Wanting more free time in 1903, he made a deal with Stevenson to be a 50/50 partner.

"At the earliest time, my grandparents lived at Congress and Preston Avenues where they had the carriage shop. This was a residential neighborhood. People would leave their carriages and horses at my grandfather's shop to be cared for.

"Most of the carriage shops in Houston were owned by one of six brothers in my grandfather's family who had moved to Houston from Colorado. They were competitors of each other."

The Stewarts came from Scotland to Maine, and from there moved to Pueblo, Colorado where they were in the blacksmithing business. When the business in Colorado was flooded out for the second time — C. Jim was about ten years old — the family migrated to Houston, where each of the seven brothers set up his own shop. The Stewarts dominated the blacksmithing business in Houston, with their horseshoeing parlors scattered on the outskirts of town, until one of C. Jim's employees by the name of Eller wanted to go into business for himself.

"Granddad gave him permission to have his own shop, (that's the way they did things in those days — you had to ask permission to leave your employer and go into business for yourself) but told him to set up shop off Leland. So that's where Eller and Grip was located," C. Jim II recalled.

As a child, C. Jim II would go with his grandfather to the prize fights held every Tuesday night behind the Sears Roebuck store building on West Gray.

"We always sat in the corner, and I got to hold the fighter's robe on my lap. Granddad was a wild man wanting to go to the fights every Tuesday night. My grandmother said he was 'unpredictable.' He didn't hesitate to get into a scrap, and loved bare knuckle boxing.

"On the other hand, he was the kindest, gentlest kind of person.

"I remember when I got my own horse, which cost me $15. Granddad didn't help me pay for it, but he blessed it as a good horse. I thought it was the greatest horse in the world, and he went along with my opinion," C. Jim II recalled with a chuckle.

C. Jim Stewart I hung pictures of his trotting horse on the walls of the horseshoeing parlor.

1899 Houston City Directory Listings

C. J. Stewart, co-founder of the company, was born in 1869. He and his wife Josephine had two children, Ross, born in 1898, and Aileen, born in 1900. According to the 1899 Houston City Directory, C. Jim was living with his young family at 2711 Harrisburg and he and Morris D. Foote were the proprietors of C.J. Stewart & Company and Houston Carriage Works at 802-804 Congress. C. Jim Stewart died in 1938.

Also in the 1899 listing, Joseph R. Stevenson, the son of John Wister and Rebecca Rosenfield Stevenson, born August 14, 1880, was living with his parents at 803 Waverly and was employed as a carriage maker for Mosehart Keller Carriage and Wagon Factory. In November 1902, he married Mary Emma McDonald. They had two children, Madlin, born in 1904, and Joseph Donald, born in 1910. Joseph R. Stevenson died in 1953.

In the 1902-1903 Houston City Directory, Joseph R. was still listed as a woodworker for Mosehart Keller, so the 1904 edition of the Southwestern Telegraph & Telephone Company Directory is possibly the first public directory listing for the new enterprise of C. Jim Stewart & Stevenson. The listing is as follows:

2892-1 STEWART, C. JIM & STEVENSON
Horseshoeing and Blacksmithing
1712 Congress Avenue

By the 1920s, the Stewart family lived at 222 Marshall and the Stevenson family resided at 804 Harold, both in the Westmoreland area of the rapidly expanding city of Houston.

C. Jim Stewart, "Co-founder" of C. Jim Stewart & Stevenson (1902 - 1938).

Joe R. Stevenson, "Co-founder" of C. Jim Stewart & Stevenson (1903 - 1953).

Madlin Stevenson: Family History

Madlin Stevenson was born in 1904, one year after her father, Joe R. Stevenson, borrowed $150 from his mother-in-law to go into business with C. Jim Stewart. (The other half of the $300 price tag for a half interest in the business was "to be payable on or before" April 4, 1904, according to the notarized contract.)

"My grandmother McDonald made him pay back the loan with interest," the beautifully coiffed, impeccably dressed, 95-year-old Ms. Stevenson recalled in a January 28, 2000 interview.

Her account of the Stevenson family history offers a unique glimpse of Houston in early 20th century.

According to Ms. Stevenson, both her father and his mother were born in Houston, but his grandmother Rosenfield (his mother's mother) came from Vienna, Austria. After arriving in Houston, her husband died and left her with five daughters to raise.

Daughter Rebecca, who became Joe R. Stevenson's mother, married a handsome, six-foot, two-inch adventurer from an aristocratic Philadelphia family. John Wister Stevenson reportedly was related to *Treasure Island* author Robert Louis Stevenson and Owen Wister, author of *The Virginian*.

Ms. Stevenson recalled that her Grandmother Stevenson lived "in the prettiest house" on West Gray.

"When I was a little girl, we would go riding on Main Street every Sunday. Papa would harness the horses—we had a surrey, and it had fringe on the top, too. We would go out Main Street to McGowen. We couldn't go any further because it wasn't paved. It was bricks, and you know how rough that is. Anyway, that was what we would do on Sunday. We used to ride all the way to McGowen and back, and that was about 10 blocks, I guess," Ms. Stevenson related.

"Papa was one of the first people to have a car in Houston. He had a five passenger Buick. I used to have a picture of Papa in the Buick with the top down. He had on a duster and his cap. General Motors wanted the picture and I gave it to them. I wish I had it back."

The Sunday rides down Main Street continued in "Papa's Buick, with the top down. It was bumpity, bumpity, bump, and if you got off Main Street, you got stuck in the mud, and we were stuck in the mud every weekend. You never saw so much mud in your life," Ms. Stevenson said.

Ms. Stevenson recalled that every Sunday her father had to fix the carbide lights on the car.

After a fire burned most of the north side of town, the Stevensons moved from Nance Street to 1207 Elgin, which was next to Kincaid School, and stayed there until 1920.

"Then Papa built a house between Montrose and Lovett on Harold Street that is still there. It's stucco and has a medallion. A real good looking house — 804 Harold is the address," Ms. Stevenson said.

While living on Elgin, the city insisted that the Stevensons get rid of the horse in the back yard.

"Mama just cried when she had to get rid of Nellie—that was the horse's name," Ms. Stevenson recalled.

According to Ms. Stevenson, the Stewarts lived over the blacksmith shop. (Address listed for C. Jim Stewart from 1908 to 1920 was 310 Dennis. The 1920-21 Morrison Fourmy Directory for Houston lists the Stewart's address as 222 Marshall.)

"I knew the Stewarts real well, particularly Ross and Katherine. Katherine was perfectly beautiful. She came from up north. Her aunts brought her down here. They lived at the Rice Hotel. She made her debut down here and then met Ross.

"I didn't know Aileen. She lived in Beaumont. Married a fellow named Tom Langham and she raised her children over there.

"I knew C. Jim Stewart real well. He was older than Papa, by about 10 years. I was real fond of Mrs. Stewart, Josephine. She was fond of opera, and I was too, so we used to go to concerts together. I used to take her as she got older.

"The Stewarts were not Houstonians, but I don't know where they came from. Colorado, I think. He and Mrs. Stewart got along real well. He and Papa got along real well, but not my brother. He was a hellion, just like Mama. Father was quiet and good, never crossed anybody. Had a wonderful disposition. He loved people," Ms. Stevenson said.

Of her generation of the Stevenson family, Ms. Stevenson said, "There was just my brother (J. Donald Stevenson, who followed Ross Stewart as president of Stewart & Stevenson) and me. You haven't lived if you haven't met my brother. He had a personality that wouldn't quit. If he liked you, he would do anything for you. If he didn't, you'd better watch out.

"Donald didn't go to college. Mama didn't make him. She didn't like school either. But he seemed to be smart enough without it. He had a real good business head. He loved to fish, and my sister-in-law would fish with him. He had the two boys, Keith and Donnie, who are just like my children because I never had any," Ms. Stevenson said.

Ms. Stevenson was married for a short time to a sea captain, who was away three months at a time. After they divorced, she studied at Parson's School of Design in New York and earned a degree from Feather and Feather. She worked as a designer at Sunniland Furniture Co. where she had a 43-year career.

Although she never worked for Stewart & Stevenson — "women didn't work when I was coming up," she said — she did recall her father's concern for the company's finances during the Depression years.

Only the Best Service and Workmanship

From the outset, C. Jim Stewart and Joe R. Stevenson determined that their business would provide only the best service and workmanship. Company stationery proudly proclaimed their respective trades:

"Horseshoeing Carefully Executed. Dealers in and Manufacturers of Fine Wagons, Phaetons, Carriages, Bike Buggies, Runabouts, Tops, Cushions, Dashes and Storm Aprons."

The "horseshoeing parlour," as it was called, was the fanciest in town. It had a wooden floor, smart wallpaper on the walls, and was so clean they located the newly hired bookkeeper's desk there. Joe carved a 16-foot-tall wooden horseshoe that hung in front of their 1712-1714 Congress Avenue shop for a sign.

The horseshoe parlor also had a pen for holding mean animals brought in to be shod. One story that contributed to C. Jim's reputation as a scrapper was that, using a "Fitzsimmons swing" and uppercut, he knocked a drayman into the street who had "cussed" him for hitting one of his mules on the leg with a hammer to get the animal to stand still. When the drayman came to, he is said to have commented, "That sure is a mean mule."

C. Jim Stewart & Stevenson hired four men mounted on fast ponies who did nothing but bring in animals needing shoes and then deliver them shod back to their owners. C. Jim and his hired smiths were soon shoeing 100 animals a day.

Joe, meanwhile, was becoming known as the best carriage maker in Houston. The rigs he made and repaired had a certain air of elegance about them that appealed to the town's wealthier citizens. A 1904 Bill of Sale to Mr. H. Masterson listed the following repairing, painting and trimming costs on November 11:

"Painting Phaeton, $18.00; 1 New Rug, $1.50; Trimming Shaft, $1.50; and Stitching Back Curtain, $.25."

That same year, Joe Stevenson and crew proudly displayed a new delivery wagon they had built for Armour Packing Company.

What was the cost of obtaining a new wagon for business from the best carriage maker in town? A January 16, 1910 page of advertisements in the *Houston Post,* which dealt mainly with wagons, buggies, harness and horseshoeing, contained this small ad:

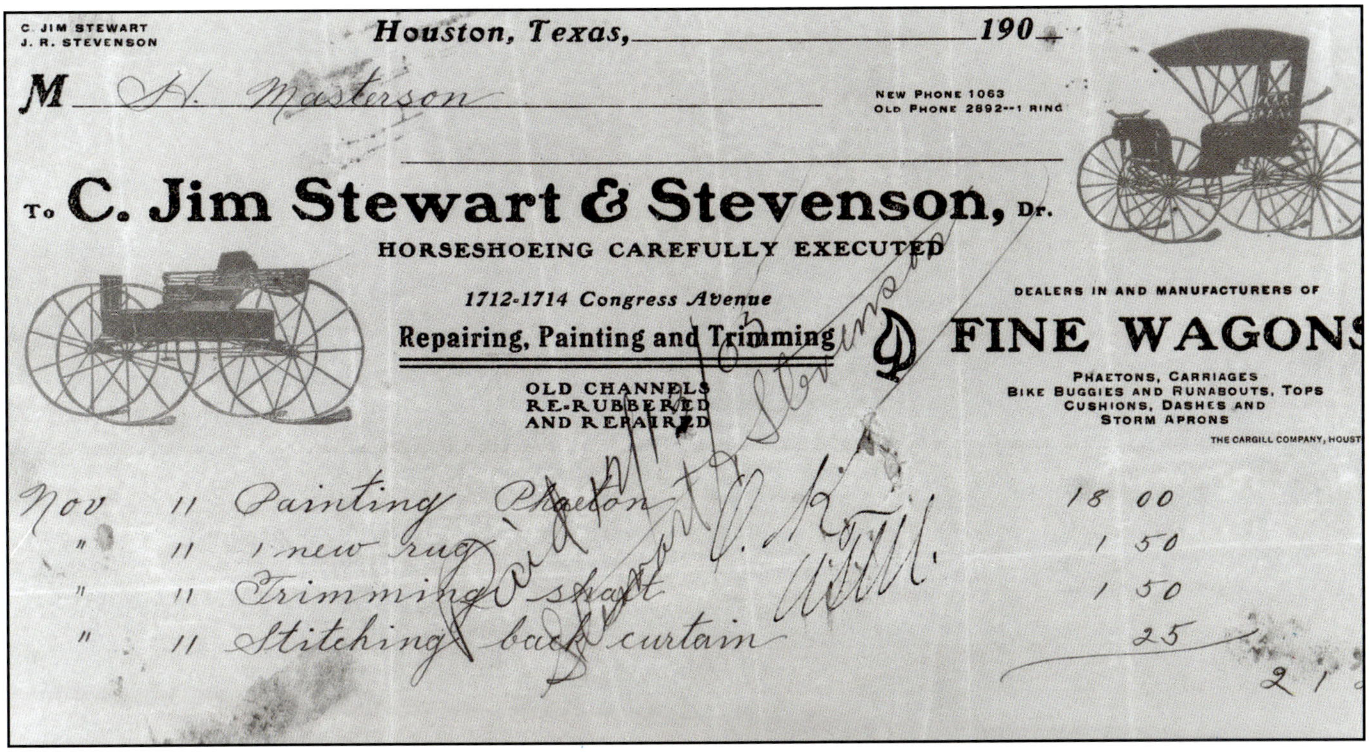

C. JIM STEWART
J. R. STEVENSON

Houston, Texas, ____________ 190__

M H. Masterson

NEW PHONE 1063
OLD PHONE 2892--1 RING

To **C. Jim Stewart & Stevenson,** Dr.

HORSESHOEING CAREFULLY EXECUTED

1712-1714 Congress Avenue

Repairing, Painting and Trimming

OLD CHANNELS
RE-RUBBERED
AND REPAIRED

DEALERS IN AND MANUFACTURERS OF

FINE WAGONS

PHAETONS, CARRIAGES
BIKE BUGGIES AND RUNABOUTS, TOPS
CUSHIONS, DASHES AND
STORM APRONS

THE CARGILL COMPANY, HOUST

Nov	11	Painting Phaeton	18 00
"	11	1 new rug	1 50
"	11	Trimming shaft	1 50
"	11	Stitching back curtain	25
			21

"Business wagons custom built $70 to $200."

The obvious pride the partnership took in its first-class workmanship was exhibited by this statement included in the "Sacrifice Sale of Fine Buggies" ad that appeared in the *Houston Daily Post* in April 1908:

"Our horseshoeing parlor is manned with exceptional skilled labor – no boys."

Two men among the shop's skilled employees were Gus Olzie and Oswald Dunbier.

C. Jim Stewart & Stevenson also took pride in the use of "scientific equipment." At a cost of $3.00, C. Jim acquired a Russell's Foot Adjuster or Leveler, manufactured in 1872 by the M&M Machine Company of Cincinnati. The latest in horseshoeing equipment, this instrument gave the correct degree needed in leveling the hoof prior to fitting the shoe. His blacksmith tools included the anvil, cross pein sledge, plain eye brad hammer, poker, hardy tool, farrier's half-round hardy, hoof parer, and farrier's horseshoeing tongs. Joe's carriagemaker's tools included a spoke shave, parallel bar clamp, level, and block plane.

One of the firm's best customers was Joe Weingarten, who conducted an active delivery trade from the small grocery he owned at Congress and Crawford. Another was Spencer Hutchens, who in 1896 had brought the first rubber-tired carriage to Houston. When Hutchens brought his rig to the shop to be serviced, Joe Stevenson took that opportunity to learn about tires. After the horseless dray became popular, he devised a process for making solid rubber tires for trucks. The Goodyear Tire & Rubber Company, his supplier of carriage rubber, then began supplying C. Jim Stewart & Stevenson rubber for tires.

Expanding to the Horseless Carriage

As C. Jim Stewart & Stevenson concentrated on shoeing horses, building and repairing wagons and buggies, the "horseless carriage" was becoming more and more popular. By 1908, automobiles were quite prominent on the streets of Houston, as elsewhere.

In 1905, C. Jim Stewart & Stevenson had its first opportunity to work on an automobile, a badly burned Dixie Flyer. The damage was so extensive that a new, wooden four-door body was handcrafted and installed. This job marked C. Jim Stewart & Stevenson's transition from horseshoes to horse power, and the beginning of the innovation and versatility that would mark the company's steady progress through its first century.

Like carriages, early automobiles were partially constructed of wood. They required similar maintenance and repair, and individual craftsmen used their skilled hands to reconstruct wood and metal parts. The partners went into the manufacture of wood parts for auto bodies and the repair of autos, trucks and truck bodies, doing a general line of repair work, painting, repairing and manufacturing springs. C. Jim used his blacksmith skills to make springs and other hardware while Joe crafted wooden wheels, axles and body parts for automobiles and trucks. One of their earliest customers was the Ford Motor Company, for which they manufactured wooden station wagon parts.

Expanding upon parts and repairs, C. Jim Stewart & Stevenson began manufacturing truck bodies for bakeries and beverage manufacturers.

As the company expanded its business, the number of employees increased. Among those assisting in this new business were Ross Stewart (C. Jim's son), S. M. Delhomme, Henry Hall, F. R. Peterson, Percy Howell, and A. B. Griff.

Marion-Handley Dealer, 1713 Preston

About 1911 or 1912, C. Jim Stewart & Stevenson became a dealer for the Marion-Handley automobile. Needing more space, the partners bought the two-story, red brick building on the corner of Congress and Chenevert, which had once housed Big Annie and her girls but had since been turned into a somewhat more respectable hotel. They lined the stairsteps to the upper floor with plastic seat cover material and Joe Stevenson turned Mustache Jim and Big Annie's former residence into an office.

Then, the firm erected a modern structure, built to meet their exact requirements, at 1713 Preston. C. Jim Stewart & Stevenson now occupied the entire eastern half of the block from Preston on the south bordering Chenevert on the east and Congress on the north.

A June 9, 1919 *Houston Chronicle* article featured a picture of the new building with the caption heading: "An Old Firm in a New Home: C. Jim Stewart & Stevenson - The House that Grows with Houston, 1713 Preston - To the Corner and Through the Back to 1712-1714 Congress." The text of the caption proclaimed: "This building, figuratively, has been under process of construction for more than twenty years. Twenty years of experience, twenty years of good will, and twenty years of honest and intelligent service to this community. Will a house resting upon such a foundation endure?"

C. Jim Stewart & Stevenson's Marion-Handley dealership on the corner of Congress and Chenevert, c. 1912.

Families Acquire Automobiles

Like everyone else in America, the Stewart and Stevenson families gradually gave up their trotting horses and carriages and acquired automobiles. Earlier records were not available, but tax receipts for the years 1923-1929 reveal that a Dodge was owned by C. Jim Stewart & Stevenson in 1923, and in 1929 the company owned a GMC Pontiaç, Dodge Touring Car, Dodge truck, and Ford Roadster.

C. Jim Stewart paid tax on a Cadillac in 1923, and he still owned a Cadillac — a touring car — in 1929. He also paid tax on a Buick and Olds Coupe in 1929.

Although J. R. Stevenson paid tax on an auto in 1923, the brand wasn't revealed on the tax records. A Hudson was listed on Stevenson's 1924 tax receipt, and his 1929 tax records included a Ford Roadster, Hudson, Ford Coupe and Buick.

Obviously, the $15 per week salary cap no longer applied — or had been increased a little.

A New Generation Comes Aboard

Ross Stewart, born in 1898, had been an eager four-year-old "assistant" when his dad opened his blacksmith shop in 1902. Working in the shop as a youth, he came to hate horses because it was his duty to clean out the stables. The lack of a necessity for that chore no doubt fueled his early love for the "horseless carriage."

Ross Stewart began work full time at C. Jim Stewart & Stevenson in 1920 after studying engineering at Texas A&M and serving in the Air Force at the end of World War I. On his return from the Air Force, Stewart saw that the automobile was becoming the mode of transportation, and he designed and built four or five cars of his own. But Fisher Body, which bought the northern blacksmith shops that were building bodies for Buicks, Oldsmobiles, Pontiacs, etc., never came South; thus C. Jim Stewart & Stevenson did not become a part of what eventually became General Motors. Instead, Ross Stewart used his engineering degree to lead the company into the repair and maintenance of automobiles and trucks, and began selling trucks and tires.

July 3, 1939

Mr. B. T. Erwin
1935 Wroxton Ave
Houston, Texas

Dear Mr.Erwin:

DON'T BE A FENDER-FLOPPER!

You can have QUICK FENDER REPAIRS at our new department. It is now possible for you to have your fenders repaired QUICKLY, BEAUTIFULLY, and INEXPENSIVELY!

Come see what we mean, or call Capitol 5311, and we will pick up and deliver your car. Get your fenders repaired QUICKLY while you see a show, while you're shopping, playing golf, or visiting with us.

Smilingly yours,

C. JIM STEWART and STEVENSON
Ross Stewart
Donald S Stevenson

The automakers developed separate dealers for each vehicle — Pontiac, Oldsmobile, Buick, etc.— and the network of new car dealers grew rapidly. But the dealers only wanted to sell new cars, they didn't want to rebuild wrecked cars. So Ross Stewart made a deal with Houston area dealers to do the body work for them. C. Jim Stewart & Stevenson stationed estimators at each of the dealers to tell the dealer how much it would cost to put a wrecked car back into shape. Then they brought the wreck to the shop on Congress Avenue, rebuilt it and sent it back to the dealer. The customer frequently never knew that C. Jim Stewart & Stevenson had made the repairs.

A dynamic force in the building of the company, Ross Stewart's initiative, dedication and tireless energy guided the company through most of its growth during C. Jim Stewart & Stevenson's formative years.

Ross Stewart married Katherine Rial in 1921 and they had two children, Nancy, born in 1922 and C. Jim II, born in Houston in 1925. Ross Stewart died in 1975.

J. Donald Stevenson, son born to the "original pardner" in 1910, joined the company in 1928. He began his career in the shop, becoming expert in appraising wrecked automobiles and working with the insurance companies. He worked closely with the shop mechanics, paint and body personnel.

During the coming World War II years, Donald Stevenson's expertise in automotive and truck maintenance and refurbishing would be called upon by the Army Ordinance Department, used to train Army personnel in the maintenance and repair of Army trucks and other vehicles. His steadfastness during the World War II years would enable the company to successfully complete military contracts on demanding schedules.

Joseph Donald Stevenson married Mildred Rose Doyle in 1932 and they had two sons, Donald Edward, born in 1944 and Keith Thomas, born in 1946. Joseph Donald Stevenson died in 1989.

Left: Ross Stewart and Donald Stevenson team up to advertise "Quick Fender Repairs" at the company's new department. The 1934 letter to Mr. Erwin begins: DON'T BE A FENDER-FLOPPER!" and the last sentence reads: "Get your fenders repaired QUICKLY while you see a show, while you're shopping, playing golf, or visiting with us."

A Growing Enterprise 1919 - 1938

More Than a Partnership, an Organization

C. Jim Stewart & Stevenson more than endured, it progressed. In its new "house," efficiency was raised to the very highest point of achievement and service increased to meet the most exacting demands. The horseshoe-shaped "C" in the C. Jim Stewart & Stevenson logo now featured an automobile, and advertised the company's new focus: Heavy Auto Accessories — Trimming, Painting, Body Building.

Efficiency must have reached a high point indeed, for with prices like these — $3.50 to paint a Ford hood, $2.00 to paint both rear fenders — it took a lot of painting and repairing to reach the $150,000 in annual revenues which C. Jim Stewart & Stevenson attained during the second decade of the new century. Before 1930, C. Jim Stewart & Stevenson was having $250,000 revenue years.

The number of company employees grew along with the company's facilities. Micky Cox headed up the automobile painting department. His assistant was Sam Loreno. Head mechanic was Fred Mitchell, who was assisted in that department by mechanics Teddy Meiselbach and Red Priest. Gus Olzie and Oswald Dunbier, holdovers from C. Jim Stewart & Stevenson's blacksmithing days, continued to work in the shop. Charlie V. Foroi came on board in 1920.

The partners discovered that some companies needed more than truck bodies and parts, painting and repair – they needed trucks. So C. Jim Stewart & Stevenson ventured into selling trucks, first representing International Trucks, about 1915. Early trucks had wheels that resembled a wagon wheel: very large in diameter and using solid rubber tires.

C. Jim Stewart & Stevenson received trucks like this from the manufacturer, then designed and built bodies to meet each customer's needs. The company had the only hydraulic press in Houston used to press the solid rubber tires onto the wheel.

C. Jim Stewart & Stevenson became a General Motors truck dealer in the early 1920s. Pictured (left to right): C. Milby Dow, J. R. Stevenson, C. Jim Stewart, O. L. Arnold, Vice President of General Motors Truck Company, Ross Stewart, and Pat E. O'Connor, assistant to the GM Truck Company president.

Only the Best Products

The partnership progressed into an organization representing different lines of merchandise. For twenty years, the firm had handled products of Goodyear Tire & Rubber Company. Now C. Jim Stewart & Stevenson represented International Motor Truck, whose product the partners declared to be the "kind of truck which fits our idea of constructive and enduring business." They custom built truck bodies for specific delivery purposes and also modified them for specialized uses, such as adding a UF Winch for use in the oil field.

In addition, they determined that: "Every commodity we sell — utility trailers, Tuthill Titanic Springs, Valentine's Varnish, Neverleek Top Material — is the very best of which we have knowledge and sold by men whom we believe will tell the truth to the public." Then, as now, Truth in Advertising was not a high priority for all companies. Not all products lived up to the promises of their spokesperson, but C. Jim Stewart and Joe Stevenson determined that anything they represented would work as promised. By the same token, they expected their customers to live up to their responsibilities. The motto of the Houston Retail Credit Men's Association, of which C. Jim Stewart & Stevenson was a member, was printed on each monthly statement and read: "Character is the Basis of Credit."

In 1923, C. Jim Stewart & Stevenson became a GMC truck dealer, and continued as such until 1936. Ross Stewart's acquaintance with Kenneth Carsey, salesman with the Cleveland Diesel Division of General Motors, would lead the company in 1938 into a totally new field, representing General Motors Diesel engines.

THE KEYWORTH DAIRY INC.
Keyworth
DAIRY Inc.
HOLSTEIN MILK FOR BABIES
PASTEURIZED MILK-CREAM-BUTTER
PRESTON 1147
HADLEY 3000
PERMIT
No. 768
Keyworth
DAIRY Inc.
PRES. 1147
HAD. 3000
KEYWORTH'S DAIRY REFRIGERATED TRUCK
HOLSTEIN MILK
FOR BABIES
AND
A-GRADE MILK
Produced and Delivered
FROM OUR FARM
PHONE
HADLEY 3000

C. Jim Stewart & Stevenson built the bodies for the trucks pictured on these two pages. *Left page, top:* Model K-72 GMC trucks circa 1920-1925. *Bottom:* Three of the Keyworth Dairy trucks (1926-28) in front of the dairy's building. Sides of the trucks, one of which is refrigerated, are painted with slogans "Pasteurized Milk - Cream - Butter" and "Holstein Milk for Babies."

This page, right: a gravel dump truck (1929-30.) *Center left:* Cook Paint & Varnish truck at its loading dock, 1928. *Center right:* The first utility truck body the company built for Houston Gas & Fuel/United Gas Corp., 1946. The body for the Ford truck was designed by J. Donald Stevenson. *Bottom:* An early trailer truck "rig" parked in front of C. Jim Stewart & Stevenson GMC Truck Co., 1932. Note that the car in the logo "C" has been updated to a more modern version than the original logo. Also note the brick streets and street car tracks.

1713-15-17-19 Preston Ave
1712-14 Congress Ave

Jim Stewart & Stevenson

HEAVY AUTO ACCESSORIES
TRIMMING, PAINTING, BODY BUILDING

Phone Pres. 434 / 435

HOUSTON, TEXAS

Retail Prices on Bake Enameling
ON CAR PARTS

These Prices Do Not Include Taking Off, Replacing or Disassembling Parts.

PRICE EACH

MAKE OF CARS	Hoods	Front Fenders	Rear Fenders	Front Splash Pans	Side Splash Pans	Head Lamps	Radiator Shells	Hood Rests
Ford	$3.50	$3.00	$2.00	....	$2.20	$1.25	....	$ .25
Cars in Class— Chevrolet, Maxwell, Overland, Etc.	4.50	4.00	3.00	1.50	2.25	1.25	2.00	.35
Cars in Class— Hudson, Buick, Dodge, Chandler, Oakland, Etc.	5.50	5.50	4.50	1.75	3.00	1.50	2.25	.50
Cars in Class— Cadillac, Lincoln, Peerless, Packard, Pierce Arrow, Marmon	6.50	6.00	5.00	2.00	3.50	2.00	2.50	.75

Side Splash Pans with Tool Boxes or Battery Boxes, each..$5.00
Radiator Shell with Shutters, each........................ 4.00

All Roadster Rear Fenders take same price as Front Fenders.

All Wire Wheels, each....................................$4.00
Bumpers, each.. 2.50

BEFORE— This is really a "serious" wreck involving wheels, axle, chassis, body and top. Looks hopeless, doesn't it? Yet...

AFTER— Here is the same car shown above. It seems hard to believe tha it was recently almost a complete wreck.

Above: Before and after pictures of a wrecked automobile repaired by C. Jim Stewart & Stevenson.

Right: a Depression era ad listing service and repair costs and urging Houston's car owners to give "this fellow a job!" Signed by Ross Stewart, the ad continues, "We are making it easy for every auto owner in Houston to break the back of the Depression at a real advantage to himself."

Ross Stewart: A Man of His Word

Earl Lester, Jr. brought his car to Stewart & Stevenson Services for repair after a collision. The repair bill was $350. He went to Ross Stewart and said, "Mr. Stewart, this is entirely too much money for what you did for me."

Stewart replied, "How much time do you think it would take to take off your bumper and radiator for repair?" Lester, a pretty good mechanic himself, replied that he could do it in two hours.

So Ross Stewart said, "If you can do that, I will mark this bill paid."

Lester went to work and removed the bumper and radiator with 15 minutes to spare.

A man of his word, Ross Stewart marked the bill paid and Lester drove off in his repaired car.

Lester later became a member of the Stewart & Stevenson Board of Directors.

(Story told by Dick Conolly, who retired as S&SS VP of Branches and Dealers Worldwide in 1975 after 30 years of service with the company.)

Operating Divisions Formed

C. Jim Stewart & Stevenson's first operating division — the Alemite Company of the Gulf Coast — also came into existence in 1923, when the company became the exclusive distributor for Alemite lubricating equipment and Stewart-Warner instruments for the automotive, industrial, farm and marine markets in Southeast Texas.

Stewart & Warner Corporation of Chicago had invented a special fitting which sealed joints. Instead of just slapping a handful of grease on a joint, mechanics now had to use a special handgun developed by Alemite to lubricate joints with the fitting. The automotive lubricating center evolved into industrial lubrication.

Hal Cone, the outside salesman for the division, went out of his way to help his customers and contributed to the company's growing reputation. By 1930, a second Alemite office was opened at 71 St. Charles Street in New Orleans, with Elmer DeBartello

C. Jim STEWART & STEVENSON Inc.

ALEMITE COMPANY
OF THE GULF COAST

A DIVISION OF
C. JIM STEWART & STEVENSON, INC.

Symbol of SW Excellence — ALEMITE DIVISION STEWART-WARNER CORPORATION

in charge of sales. Tax records for 1934 indicate that C. Jim Stewart & Stevenson of 816 St. Charles Street, represented by Wallace A. Farrand, purchased a used 1933 Ford sedan for the sum of $34.50 from Homan L. Sproule. The Alemite salesman had new wheels.

By 1938, the Alemite Division was a major part of the company's business, with operations all along the Texas and Louisiana Gulf Coast. Bob Hardy, who started with the company as credit manager, was promoted to manager of the Alemite Division.

C. Jim Stewart & Stevenson Distributing Company was formed in 1924 when the DuPont organization selected the company as the exclusive distributor for its new Duco automotive enamel paint line. Before, automobiles had been painted with a brush. Ross Stewart took on a franchise from DuPont for the Houston to New Orleans territory. He began training classes to teach spray painting techniques, and he and others traveled to body shops all over to teach people how to use the spray paint system, as well as the Alemite lubricating system.

In addition to the DuPont paint franchise, other products related to the automotive supply line were added and handled through the Automotive Paint & Supply Division of C. Jim Stewart & Stevenson. Abbie Curneal was the outside salesman for this division, which was a dealership for Gillbarco service-station gasoline pumps, various kinds of automotive paint and trim supplies, small hand tools, jacks, lug wrenches, A.C. spark plugs, Prestone anti-freeze, etc. It served as sort of a supermarket or wholesale house to other automotive and truck repair shops and dealerships who were in the business of repairing automobiles and trucks, as well as painting them. Charlie Foroi, the main person behind the counter, was always on the job early, cheerfully greeting customers who came in for paint and trim supplies. The company became DuPont's largest dealer for automotive paint and supplies.

Shortly after Prohibition ended in 1933, Hertz Systems, Inc. came to Ross Stewart and said they needed a Hertz truck franchise in the Houston area to handle an account for Grand Prize Brewery, which was

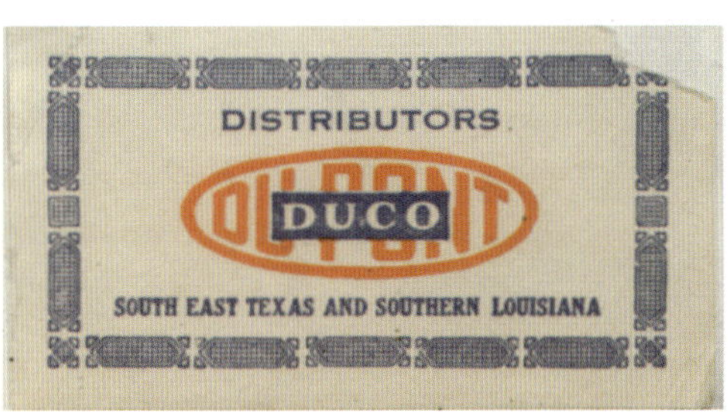

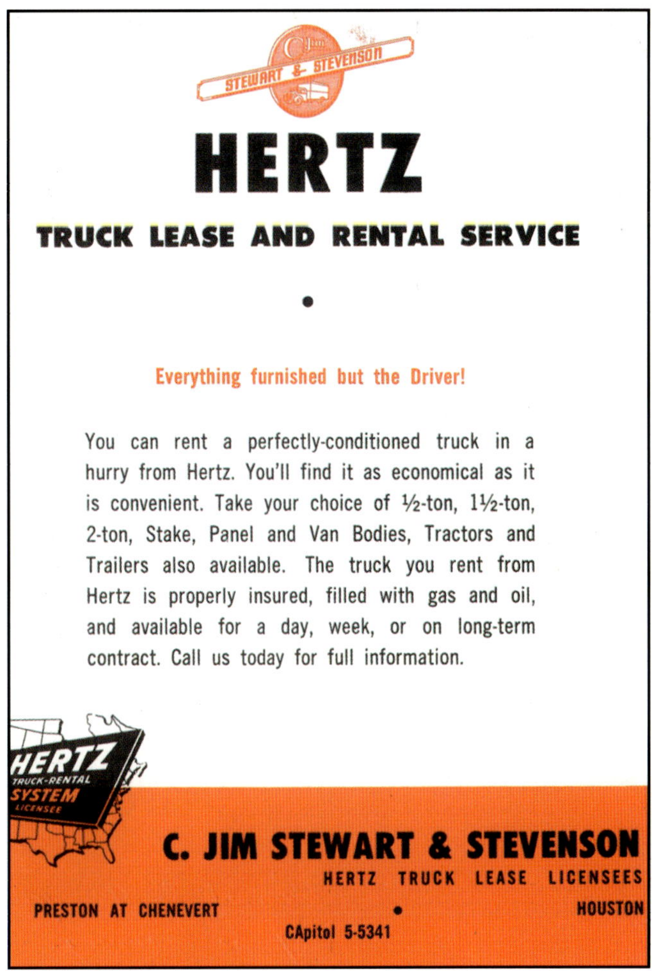

THE QUESTION OF--- "PRICE"

C. Jim Stewart & Stevenson enjoys a very wide reputation for the high character of work performed by this organization, and the superior excellence of the service it renders.

Because of this

enviable reputation it might be supposed that the prices charged by us must necessarily be high.

We wish to dispel any such notion.

For any type of work indicated or desired, our prices compare favorably with those of any other first-class organization operating in this territory.

Moreover, because of

our wide facilities for doing any character of job, we are enabled to give you exactly the type of work you want.

Obviously, a cheap car, or one of old make, would not justify the expenditure that an expensive one of recent model would. In such a case, service and utility are more important than precision and appearance.

And obviously again

the prices that would have to be charged would be less than where precision and fine finish are demanded.

We are equipped and qualified to perform the very highest grade of workmanship available anywhere. But we are also equipped and qualified to give exactly the type of workmanship demanded by individual requirements.

And our prices, in every instance, will be found fair and equitable and in line with those of competitors.

Prove this for yourself by giving us a trial on that next job.

Build good will and increase your profits by making use of the facilities which we have here in our establishment. Your inquiries shall receive prompt, courteous attention.

C. JIM STEWART & STEVENSON

Heavy Auto Accessories, Trimming, Painting, Body-Building

HOUSTON, TEXAS - Phone Preston 0434
1713-15-17-19 PRESTON AVE., 1712-14 CONGRESS AVE.

C. Jim Stewart & Stevenson employees and their families prepare for the annual company picnic in this 1920s era photo. The company picnic is an annual tradition.

owned by Hughes Tool. C. Jim Stewart & Stevenson took the franchise and bought trucks, built special bodies for them, and leased them to Grand Prize, the first customer for the company's fourth division, the C. Jim-Hertz Lease Truck Division formed in 1937. The fleet of trucks grew as the company went more and more into truck building. This division began its steady growth into one of the major truck leasing operations in the South, eventually peaking with a fleet of 500 trucks. The division was managed by George Felts.

All of these divisions — Alemite, Distributing, Automotive Supply, Hertz Truck Lease — were begun by Ross Stewart, who was always on the alert for new products, new ideas, and new directions to take the company. Each of these divisions became a profit center contributing to the overall profit of the company.

A Coca-Cola delivery body built by C. Jim Stewart & Stevenson on a Chevrolet/GMC truck, about 1946.

The Pain of Expansion

In 1972, Ross Stewart, reminiscing about the company's history as it expanded during the 1920s, said, "I put these places in, personally. I was home very little. Katie (Mrs. Ross Stewart) says, '...you didn't know your own children, hardly...' while I was out beating on those doors. I used to ride around in a 4-cylinder Dodge car, and by-God, that was the roughest ride in the world. I'd ride that bastard, those roads were rough, and I would get out and stand with one foot on the running board, with one hand on the steering wheel, and have the door open and hold onto that so I could keep going. My butt would be so sore and my back would be breaking, and that way I could keep going — oh, it was awful, Goddamn, it was rough! That was when we'd go to New Orleans or Mississippi and I was working all these shops, selling Duco, Alemite, etc. At any rate, that was a period when business was pretty good."

The Diesel Engine 1938-1941

The Diesel Era Begins

With the death of C. Jim Stewart in 1938, Ross Stewart assumed leadership over all company operations and became the company's principal contact with major suppliers. One of those contacts was Kenneth Carsey, a salesman for General Motors Cleveland Diesel Engine Division, who introduced Ross Stewart to GM's new Series 71 Diesel engine.

General Motors had developed its new two-stroke Diesel engine to provide power for machinery in rural areas without electricity, and the company had asked its salesmen to recommend distributors for the new Diesel engine in their territories. Kenneth Carsey recommended Stewart & Stevenson.

(According to the *Texas Almanac*, the first rural electrification in the nation took place in Bartlett, Texas, in 1933. REA, the Rural Electrification Authority, a part of President Franklin D. Roosevelt's New Deal established in 1935, would take decades to completely "electrify" the nation's rural areas.)

Fascinated by GM's new Diesel engine, and with the imagination to see that it could be applied in numerous ways, Ross Stewart sought and won a Diesel engine distributorship for C. Jim Stewart & Stevenson. In the fall of 1938, General Motors granted C. Jim Stewart & Stevenson a franchise to distribute its new Series 71 GM Diesel engine in the Houston trade area, which included about a 125-mile radius around Houston.

"We take pleasure in welcoming you to the fold of the Diesel Engine Division of General Motors Sales Corporation," read the letter sent from the Cleveland, Ohio facility on October 10, 1938.

C. Jim Stewart & Stevenson was one of four companies granted Detroit Diesel engine distributorships in Texas. The others were Milam Chevrolet in San Antonio, Plains Machinery in the Panhandle and one other in Odessa. North Texas and the eastern part of the state were handled by Diesel Power Co., a distributor out of Oklahoma.

The Series 71 Detroit Diesel engine line consisted of a single-cylinder Model 1-71, a three-cylinder Model 3-71, a four-cylinder Model 4-71 and a six-cylinder Model 6-71. All of the engines were rated at 15 horsepower per cylinder, and 1200rpm for continuous duty. At that time, there wasn't a single Series 71 engine in the state of Texas. But that was about to change.

In 1938 C. Jim Stewart & Stevenson hired 19-year-old Joe Manning to be the Diesel engine salesman. Manning had been attending East Texas State Teachers College in Commerce, Texas, but dropped out of college after his first year to take the job.

C. Jim Stewart & Stevenson features GM Diesels at the 1939 Houston Oil Show, the first held in Houston. It later became the Tulsa Oil Show.

Left page: Two Model 1-71 GM Detroit Diesels driving a single 20kW generator.

John M. Carsey (seated left center behind table) was honored at the company's 50th anniversary sales meeting in 1952. Joe Manning (in white shirt) is seated beside him. A young C. Jim Stewart II is seated second from left.

"Ross Stewart hired me strictly on the basis of my grandfather's reputation in the Diesel engine business," Manning would later say.

Manning's grandfather, John M. Carsey, was retired from Fairbanks-Morse Company, the leading manufacturer at that time of heavy duty Diesel engines. His son, Kenneth Carsey, who was Joe Manning's uncle, was the GM Detroit Diesel salesman who had suggested to Ross Stewart that C. Jim Stewart & Stevenson become a Diesel engine distributor.

"My grandfather, whose sales had virtually built Fairbanks-Morse, was a legend in the Diesel engine business," Manning said. "Mr. Stewart's idea in hiring me, a young kid with no experience whatsoever in the Diesel engine business, was that he would automatically get the services and counsel of my grandfather."

John Carsey, who had sold Fairbanks-Morse generator sets to municipalities throughout the Southwest, did share his unique knowledge of the Diesel engine business with his young grandson, and made an indelible mark on C. Jim Stewart & Stevenson's beginning in the Diesel engine business. Among other things, Carsey was responsible for the contract form used by C. Jim Stewart & Stevenson's newly formed Engine Division. The contract contained the C. Jim Stewart & Stevenson "Guarantee of Duty," which promised that the engine could be made to perform in accordance with what the salesman had told the customer — and if it failed and the company was unable to make it perform as such, that C. Jim Stewart & Stevenson would remove it at its expense and refund any monies paid.

A Complete Parts Inventory

It was also on John Carsey's advice that C. Jim Stewart & Stevenson stocked a complete inventory of engine parts along with the Diesel engines. This consisted not just of fast-moving parts, but also cylinder blocks, cylinder heads, crankshafts and every single part necessary to actually build any of the four engine sizes. This unprecedented move to stock a complete inventory of parts enabled the company to capitalize off the competition's weakness, for when an engine broke down, it would sometimes take months to get parts from the factory. The delay would create a total disaster for a customer that depended on his engine to make a living. By recognizing that "engines are no better than the parts and service behind them," C. Jim Stewart & Stevenson could guarantee its customers that any failed engine would be put back in running condition in the shortest possible time.

Expert mechanic Fred Mitchell became the Service Department of C. Jim Stewart & Stevenson's new Engine Division, aided by mechanic Teddy Meiselbach, when needed. Parts man George Haltham soon joined the team.

Left: This small power plant, sold by Manning in 1939, provided power for the Texas Theater and offices of J.G. Long in Bay City, Texas. It was designed so that either engine could power the generator in an emergency.

Selling to a Limited Market

Novice salesman Joe Manning packed his leather briefcase with literature, competitive engine information, contract forms and files on prospects, and headed out into the countryside. Leaving the office every Monday, he worked his way by auto through the territory, stopping in one small town after the other, at farms and ranches, talking to cotton gin owners, rice farmers, dragline operators, feed mill operators, sawmill operators, theaters, hotels, cafes and any other operation that could be a potential user of Diesel engines.

His first sale was for a pair of single-cylinder, 10kW generator sets to the Scanlan Estate, which owned a vast plantation along the Brazos River 20 miles south of Houston. The two generator sets furnished the electrical prime power for a very large, three-story mansion.

The sale of a prime power plant, consisting of three 40kW generator sets for the then "very plush Wilshire Village Apartment complex" being built by Herman Cohen on West Alabama Street in Houston, was considered a major breakthrough.

Two of these Model 1-71, 10kW units supplied electric power for the Scanlan Estate Plantation Mansion located 20 miles south of Houston.

Guarantee of Duty

C. Jim Stewart & Stevenson's Guarantee of Duty was unique and expansive.

"If a customer signed for an order with us," Manning said, "our Guarantee of Duty said the engine would be suitable for the customer's application in producing the necessary horsepower needed to successfully operate the customer's machinery, cotton gin or whatever. On the condition that it failed, and we couldn't make it work, we would remove it at our expense and refund any monies paid. We also guaranteed the fuel consumption. No other company did this. If, on rare occasions, we made a mistake and to make good on our Guarantee of Duty had to take an engine back and refund the customer's money, we did so without any argument or fuss. When dealing with us, the customer didn't need any lawyers because we were always fair and honest and we made sure that the customer got a full measure for the dollars spent."

Doing Whatever It Takes

One of the Engine Division's original strengths, in addition to having all parts in inventory, was to sell a complete "turnkey" job, which included installation of the engine and all associated equipment needed to make it work.

Calling him the "Father of the Service Department," Joe Manning said Fred Mitchell was an extremely important asset to the company.

Cotton gins were among the early users of the Diesel engines.

The R. A. Fenton sawmill turns trees into lumber with Diesel power.

Owen Metcalf

Owen Metcalf joined the Engine Division in 1940, just in time to help install five 60kW GM Detroit Diesel generator sets in the basement of the Ben Milam Hotel in Houston. The units served as prime power for the hotel. When World War II broke out, Metcalf was put in charge of the Engine Division's fabrication and assembly department. After the war he was made factory superintendent.

"If a customer had a problem with an engine or power plant, Fred would get up in the middle of the night, pack the needed parts in the back of a 1935 model sedan, go to the job and take care of it.

"Fred could single-handedly overhaul a Model 1-71 engine in four and one-half hours," Manning recalled, "and he would stay on the job 24 hours a day until the problem was corrected. Fred Mitchell set that example which all customer service people since have followed. Today, the company is well known for that kind of service people."

The Engine Division's small crew was the sales force, the installation department, parts and service departments, and also made the engineering drawings — "crude though they were," Manning described them — that sometimes were required.

Customer by customer, this small Engine Division persevered, selling engines, installing them and making friends for the company. Soon word had gotten around about C. Jim Stewart & Stevenson's capabilities in the Diesel engine business. Before the 1930s decade ended, C. Jim Stewart & Stevenson reached its first $500,000 revenue year.

The Normana Theater was among early customers for the Diesel-powered generator.

World War II
1941 - 1945

War Changes Company's Direction

The bombing of Pearl Harbor on December 7, 1941 brought the United States into World War II and a swift change in C. Jim Stewart & Stevenson's course of direction.

The Diesel engine was extremely important to the war effort, and the production capacity of Diesel engine manufacturers — the Detroit Diesel Engine Division of General Motors, Caterpillar, Fairbanks-Morse & Company, the Electro-Motive Division of General Motors, Cleveland Diesel Engine Division of General Motors, Cummins Engine Company, Atlas Imperial Engine Company, Nordburg, Enterprise and a few others — would be designated to support the military effort. The production of Diesel engines by these companies came under the control of the War Production Board, which had been established to monitor and govern the production of war supplies.

The government froze all sales of Diesel engines except to certain essential industries, ultimately shutting down Diesel sales to just about everything except military services and farmers.

"We couldn't get Diesel engines during the war," Joe Manning recalled. "All were on allocation to the military, which took priority over everything that we sold engines for. Agriculture, essential to the feeding of the troops, could occasionally get one."

So C. Jim Stewart & Stevenson, conscientious about the needs of its Diesel engine customers and knowing that engines the company had sold would eventually break down and need parts, decided to go after some of the government business in order to continue having enough parts to supply its engines in the field. Since the company could get no engines to sell commercially, C. Jim Stewart & Stevenson also needed the government business to survive.

First Government Contract

In 1942, C. Jim Stewart & Stevenson obtained its first government contract.

"I was in Memphis, Tennessee on my honeymoon," Manning said, "when Mr. Ross Stewart called me and told me to go to Washington. The Treasury Department was seeking to purchase under the Lend-Lease Program some trailer-mounted Diesel generator sets for our allies, the Russian Army."

Under the Lend-Lease Program, the U. S. government loaned Allies money, which they used to lease equipment produced in the United States that was needed for the war effort.

So Manning took the train to Washington, D.C.

"When I arrived at the Procurement Division of the

Testing the 35kW mobile electric power station with the M-4 Fuel, December 1942.

Generator sets for the Russian army are crated and labeled for shipping.

Treasury Department, I was told that the engines for this requirement had to be capable of running on Russian M-4 fuel oil. This M-4 Russian fuel was very heavy and about the consistency of 160-weight differential oil. On top of that, the engines would be operating in an extremely cold climate. The procurement officer and the Russian engineer, who was making the decision, did not believe that the Model 4-71 GM Detroit Diesel engine that we were offering could possibly burn the Russian M-4 fuel successfully.

"We persisted that the GM Detroit Diesel would run on the M-4 Russian fuel," Manning said. "My grandfather said it would if we heated the fuel and got the viscosity low enough to get it through the injectors and into the cylinders, and I had no reason not to believe him. We told the people in Washington that we would simulate the Russian fuel and prove to them that the GM Diesel engine could successfully operate on the M-4 fuel."

So the Russians provided the specifications for the fuel and E. W. Saybolt Company in Houston reproduced it for Stewart & Stevenson for testing purposes. Meanwhile, C. Jim Stewart & Stevenson had modified the basic Model 4-71 GM Diesel by replacing the standard fuel filter system with a primary, edge-type duplex fuel filter and a secondary Fuller's Earth filter. Other modifications included a heating system for the main fuel tank. The company then set up a test unit in the old Congress Avenue building and demonstrated to the Russian engineer, Nickoli Vasiliev, that the engine would run efficiently on the Russian fuel. After the test, the engine was dismantled to show there were no detrimental effects from running it on this heavy fuel.

C. Jim Stewart & Stevenson got the order.

The original military purchase was for 50 mobile trailer-mounted Diesel generator sets of 35 kilowatts each, at 1500 rpm, 50-hertz. The success of those

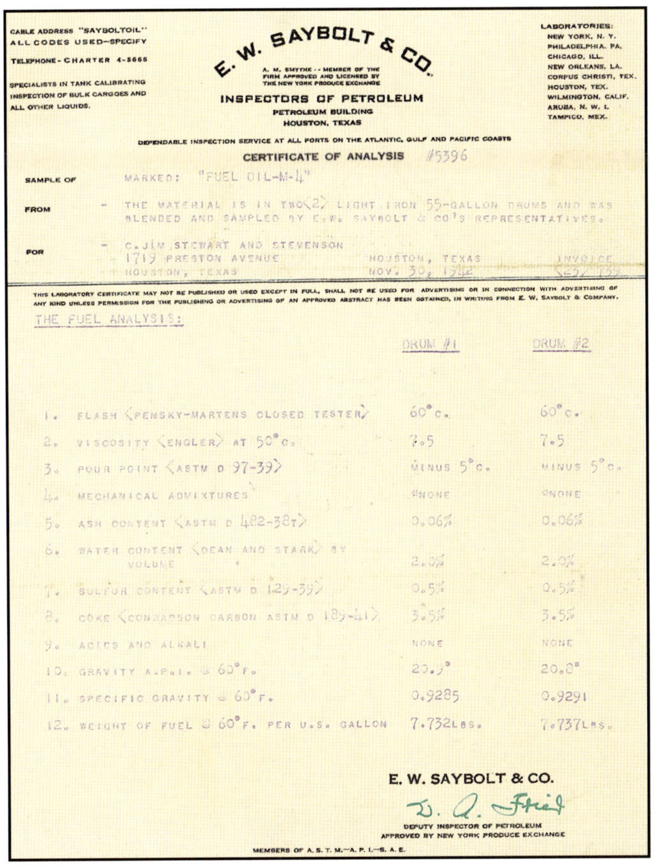

CABLE ADDRESS "SAYBOLTOIL"
ALL CODES USED—SPECIFY
TELEPHONE - CHARTER 4-5665
SPECIALISTS IN TANK CALIBRATING INSPECTION OF BULK CARGOES AND ALL OTHER LIQUIDS.

E. W. SAYBOLT & CO.
A. M. SMYTHE - - MEMBER OF THE FIRM APPROVED AND LICENSED BY THE NEW YORK PRODUCE EXCHANGE
INSPECTORS OF PETROLEUM
PETROLEUM BUILDING
HOUSTON, TEXAS

LABORATORIES:
NEW YORK, N. Y.
PHILADELPHIA, PA.
CHICAGO, ILL.
NEW ORLEANS, LA.
CORPUS CHRISTI, TEX.
HOUSTON, TEX.
WILMINGTON, CALIF.
ARUBA, N. W. I.
TAMPICO, MEX.

DEPENDABLE INSPECTION SERVICE AT ALL PORTS ON THE ATLANTIC, GULF AND PACIFIC COASTS

CERTIFICATE OF ANALYSIS #5396

SAMPLE OF MARKED: "FUEL OIL-M-4"

FROM - THE MATERIAL IS IN TWO (2) LIGHT IRON 55-GALLON DRUMS AND WAS BLENDED AND SAMPLED BY E. W. SAYBOLT & CO'S REPRESENTATIVES.

FOR - C. JIM STEWART AND STEVENSON
1719 PRESTON AVENUE HOUSTON, TEXAS INVOICE
HOUSTON, TEXAS NOV. 30, 1942 [illegible]

THIS LABORATORY CERTIFICATE MAY NOT BE PUBLISHED OR USED EXCEPT IN FULL, SHALL NOT BE USED FOR ADVERTISING OR IN CONNECTION WITH ADVERTISING OF ANY KIND UNLESS PERMISSION FOR THE PUBLISHING OR ADVERTISING OF AN APPROVED ABSTRACT HAS BEEN OBTAINED, IN WRITING FROM E. W. SAYBOLT & COMPANY.

THE FUEL ANALYSIS:

	DRUM #1	DRUM #2
1. FLASH (PENSKY-MARTENS CLOSED TESTER)	60° C.	60° C.
2. VISCOSITY (ENGLER) AT 50° C.	7.5	7.5
3. POUR POINT (ASTM D 97-39)	MINUS 5° C.	MINUS 5° C.
4. MECHANICAL ADMIXTURES	*NONE	*NONE
5. ASH CONTENT (ASTM D 482-38T)	0.06%	0.06%
6. WATER CONTENT (DEAN AND STARK) BY VOLUME	2.0%	2.0%
7. SULFUR CONTENT (ASTM D 129-39)	0.5%	0.5%
8. COKE (CONRADSON CARBON ASTM D 189-41)	3.5%	3.5%
9. ACIDS AND ALKALI	NONE	NONE
10. GRAVITY A.P.I. @ 60° F.	20.9°	20.8°
11. SPECIFIC GRAVITY @ 60° F.	0.9285	0.9291
12. WEIGHT OF FUEL @ 60° F. PER U.S. GALLON	7.732 LBS.	7.737 LBS.

E. W. SAYBOLT & CO.
D. A. Fried
DEPUTY INSPECTOR OF PETROLEUM APPROVED BY NEW YORK PRODUCE EXCHANGE
MEMBERS OF A. S. T. M.—A. P. I.—S. A. E.

The M-4 Fuel Analysis provided by E. W. Saybolt & Co.

Mobile generator sets for the Russian Army are parked along the street awaiting testing and shipment.

Generator sets, *left,* are "tractored" out of the assembly area, and, *right,* are being used by military units in the field.

units led to more contracts until C. Jim Stewart & Stevenson's Engine Division, by the end of the war, had become the nation's second largest supplier of Diesel generator equipment under the Lend-Lease system. Toward the end of the war, the Engine Division was producing, testing and shipping six 75kW trailer mounted Model 6-71 powered Mobile Power Stations each day. After testing, the units, each about half as big as a boxcar, sat along the curbs on Congress Avenue awaiting shipment. Most of this equipment went to wartime ally Russia, but some were used in D-Day landings at Normandy and in the North Africa campaign.

A new corporation, Stewart & Stevenson Services, was set up to perform the army tank engine refurbishment contract. Donald Stevenson, put in charge of the contract, had a building erected and hired women to perform light assembly work.

Refurbishing General Sherman Army Tank Engines

During the war, Federal Machine and Welding Company in Warren, Ohio, was given a contract to refurbish 800 General Sherman tanks, which were powered with the GM Detroit Diesel Twin 6-71 engines. Federal Machine, basically a welding company, had no experience rebuilding engines, so officials at General Motors Detroit Diesel advised Federal Machine to subcontract the engine rebuilding to C. Jim Stewart & Stevenson. The company accepted and fulfilled the subcontract.

Later, thousands more of the General Sherman tanks had to be refurbished and outfitted with new and heavier armor-plate and bigger guns. The War Production Board wanted General Motors Detroit Diesel to rebuild 4,000 engines at the rate of 40 each day.

"As far as the government was concerned, the General Motors Detroit Diesel Division was the only company in the country with the capability to rebuild that many Diesel engines as fast as the government wanted them," Manning said, "but GM had its hands full building new engines for the war effort.

"General Motors officials called Mr. Ross Stewart on the phone and told him that the government had this large quantity of Army tanks with engines that had to be rebuilt, and insisted that Stewart & Stevenson rebuild the units under a subcontract from GM. Mr. Ross Stewart informed them that we, too, were working seven days a week, night and day, producing everything that we could for the war effort, and that furthermore, we had no facility to rebuild that quantity of units."

When GM insisted that C. Jim Stewart & Stevenson take on the job, Donald Stevenson, son of original "pardner" Joe R. Stevenson, and Joe Manning traveled

Ruth Hill, welder hired by Stewart & Stevenson during World War II, is surrounded by jeep parts.

to the Red River Arsenal at Texarkana to look at some of the engines to see what would be required in rebuilding them.

A new corporation, Stewart & Stevenson Services, Inc., was set up and took the contract. Donald Stevenson, put in charge of the project, had a new building erected on Maplewood (now Lockwood) Street and staffed it with personnel in less than 60 days, hiring women to do the lighter work.

Using the company's previous experience rebuilding these same Diesel engines for Federal Machine & Welding Company, Stewart & Stevenson Services based its contract on a fixed price of $1900 per tank unit. General Motors officials quickly told Stewart & Stevenson its price was too low, but the company stood by its contract.

"We met the demand," Manning said. "They shipped us 40 engines every day, and every day we shipped 40 rebuilt engines back."

Not only did Stewart & Stevenson Services complete the contract on time at the price originally quoted, but also returned, unsolicited, approximately one million dollars, making sure that this customer, too, got a full, fair measure for the dollars spent.

The Harrisburg plant, originally constructed in the 1920s as a wagon factory, provided space for Stewart & Stevenson's massive war effort.

Overhauling Army Trucks, Jeeps

Another war-time project managed by Donald Stevenson was the overhaul of 2,200 U.S. Army jeeps, which Stewart & Stevenson was able to turn out faster than other, much larger companies. Years after the war, the company would hear from people — some as far away as Australia — who were still using jeeps remanufactured by Stewart & Stevenson during the war. This project also was accomplished at the Harrisburg (now Lockwood) plant.

Remanufactured Army jeeps are readied for shipping at the Harrisburg plant in 1944. *Inset:* The original Engine Division sign at the plant on Harrisburg Boulevard, 1945.

Manufacturing Machine Gun Barrel Covers

During the war, C. Jim Stewart & Stevenson continued to repair wrecked autos and paint cars and trucks. A division of that company, C. Jim Stewart & Stevenson Distributing, which sold all kinds of auto accessories, paint and trim products, and Gillbarco gasoline pumps, had a department that also made seat covers for cars. That department, located on the second floor of the old "C. Jim Building" at 1719 Preston, was converted to manufacture machine gun barrel covers out of heavy canvas material. From 1943 to 1945, the company produced thousands of these gun barrel covers, which prevented corrosion of the machine gun barrel when the gun was not in use.

A war-time employee sews machine gun barrel covers at the Preston Street location in May 1944.

C. Jim Stewart & Stevenson received many letters of commendation for its World War II efforts. Among them were letters from Commanding General Brehon Somervell, Army Armed Forces; Lt. Gen. L. H. Campbell, Jr., Chief of Ordnance, War Department; 1st Lt. Stanley Wallach, Army Airways Communications System at Ellington Field, Texas; and William S. Vensen, U.S. Army Ordnance Representative, Army Service Forces, Detroit Ordnance District.

The "C" Award for Cooperation was presented to employees, suppliers and others who contributed to the company's successful completion of mobile Diesel power stations, and C. Jim Stewart & Stevenson received a Certificate of Appreciation from Army Service Forces, St. Louis Ordnance District as a member of the Industry-Ordnance team making V-E and V-J days possible.

From War to Peace 1945 -1950

The Partnership Becomes Four Companies

After World War II ended, the original partnership became four companies: C. Jim Stewart & Stevenson, Inc. (the original C. Jim Stewart & Stevenson), Stewart & Stevenson Services, Inc. (formerly the Engine Division, which produced 80 percent of the company's product), Stewart & Stevenson Distributing Company, and Alemite Company of the Gulf Coast.

As World War II progressed, cars became in short supply and the new car dealers began rebuilding wrecked automobiles themselves to have something to do. At that point, C. Jim Stewart & Stevenson, Inc. began phasing out of auto wreck rebuilding and concentrated on its truck business.

The Hertz leasing division of C. Jim Stewart & Stevenson, Inc., at the 1718 Congress location, grew steadily after the war, serving many well-known Houston firms with large fleets ranging from the smallest pickups and panel trucks to the largest tractors and trailers. A diversified transient fleet was available to the general public by the hour, day, week or month.

Stewart & Stevenson Distributing Company, the exclusive distributor for the Duco automobile enamel paint line, now had branches in Galveston and Port Arthur in addition to the 1713 Preston location.

Alemite Company of the Gulf Coast continued as the exclusive distributor for Alemite lubricating equipment for automotive, industrial and farm use in southeast Texas, Louisiana and southern Mississippi, with offices at 1718 Congress in Houston and in New Orleans.

Prior to the war, Stewart & Stevenson Services, Inc. (the former Engine Division) had successfully utilized Diesel engines for more and more applications on a "turnkey" basis, responsible not only for supplying Diesel engine power, but for building the complete application and guaranteeing the output of the package. That success, coupled with the company's tremendous war effort, led General Motors Detroit Diesel after the war to cancel all of its other distributorships in Texas and turn them over to Stewart & Stevenson Services. A small part of New Mexico also was added to the company's territory.

After the war, C. Jim Stewart & Stevenson phased out of auto wreck rebuilding and concentrated on its truck business. This was one of the early metal truck bodies built by the company.

A young C. Jim Stewart II (front, second from left) sits with (l-r) his father, Ross Stewart, John Carsey, Arch Campbell and Bob Hunter from GM Detroit Diesel, L. O. Carlson, and Joe Manning at this 1957 sales meeting. Tom Langham is seated behind Bob Hunter. Jake Conine is wearing the cowboy hat.

Third Generation Stewarts Join Firm

C. Jim Stewart II, son of Ross Stewart and Catherine Rial Stewart, began his full-time career at Stewart & Stevenson in 1947. This third generation Stewart had served in the United States Army Air Corps from 1943 to 1945, had attended the University of Texas, and was newly married to Gretchen Elizabeth (Gay) Braun.

As a teenager, Stewart had worked most of his summers and Saturdays in the shop, straightening car fenders for repair and doing other assigned jobs.

"We were always paid at noon on Saturday," he recalled years later. One Saturday while "shooting the breeze" with some of the mechanics, they "happened" to see what he had been paid.

"'Jim, you are worth a whole lot more than that. Look what I get,' one of the mechanics told me, which was about five times what I received," Stewart said.

"'You should tell your Dad you're going to quit if you don't get paid more,'" he recalled the mechanic egging him on. So riding home with his Dad, he told him that.

"Well, Jim," Stewart recalled his father saying, "I understand how you feel. But you don't have to do that, you're fired!"

Stewart's mother was very upset about what had happened, but his father refused to back down. Instead, he got a friend at the Light & Power Company to hire Jim for the rest of the summer.

"I learned never to question what I was being paid," Stewart said.

Stewart began his full-time career at the company overseeing the Alemite and Automotive Supply Divisions, and also was put in charge of purchasing materials for the whole corporation.

"Dad wanted me to become purchasing agent for the whole company, to take the checkbook away from everyone and get control of purchasing. I had a bit of a problem, because some people didn't like that I took away their authority to buy whatever they wanted. That was my job for a long time," he said.

One of the first jobs Ross Stewart gave C. Jim II after he came to work full time was to collect a bill from a customer on Navigation Drive. Arriving at the company, he told the owner that he was Jim Stewart, there to collect money owed to Stewart & Stevenson. The guy pulled a gun on Jim, said he didn't owe Stewart & Stevenson any money and told Jim to get off his property.

Jim went to a phone, called his father and told him what had happened. Ross Stewart is said to have muttered, "You can't send a boy to do a man's job."

Taking Jim with him, Ross Stewart went to the company and told the owner, "You owe us money. Either pay it or shoot us!" He got paid.

Thomas H. "Tom" Langham III, son of Aileen Stewart Langham, also joined the company after the war, working first in the Engine Division Parts Department. He later moved to Waco, Texas as resident salesman for the Central Texas territory, where his great sense of humor and friendly personality proved enormously beneficial attributes for selling Diesel engines for cotton gins, dragline operations and irrigation projects. Having proven highly successful in sales, he was moved from Waco to Odessa to handle Diesel engine and irrigation equipment sales for the new West Texas branch operation where Jake Conine was branch manager.

When the credit manager's position became open, Tom moved back to Houston to assume that position, then became part of the management team over the manufacturing and engineering departments. As vice president, he played an active role in the development of the company's new airline ground support equipment business.

S&SS Builds on the Diesel

By war's end, Stewart & Stevenson Services had become a large organization with a great number of employees — and no commercial business. What direction should it take next?

The decision was made to take up right where the Engine Division had left off at the war's beginning: expanding the Diesel engine business for commercial purposes.

Stewart & Stevenson Services, Inc., reorganized for greater production, began selling and servicing Diesel engines as well as fabricating and manufacturing engine-driven packages: generator sets, irrigation pumping units, commercial trucks — anything that could utilize a Diesel engine as a prime mover.

The S&SS headquarters remained at the Harrisburg Boulevard location, which served as the central fabricating and manufacturing facility, as well as Diesel engine and Diesel-driven equipment sales and service for the local Houston territory. Joe Manning continued to serve as general manager, responsible for all operating functions, including sales, service, after-market parts and manufacturing. Other employees in the Stewart & Stevenson Services Houston operation at the end of the war were Margaret Cain, secretary; Charles E. "Charlie" Ames, Engine Sales; Fred Mitchell, general service manager; I.C. "Bill" McWhirter, general parts manager; J.M. "Pappy" Lawrence, operations manager, and L.O. Carlson, controller.

Margaret Cain

Margaret Cain, who joined the Engine Division as secretary in 1945, was noted for "spending 24 consecutive hours typing proposals and correspondence to meet deadlines" and for her trademark hat, which she never appeared without. Joe Manning recalled that she must have had 25 or 30 hats because "it seemed she wore a different one every day." At her retirement in 1970 after 25 years of service, she was praised for "her loyalty, faithful dedication, and unselfish attitude." Co-workers deluged her with gifts of hats.

S&S Branch Offices Established

Stewart & Stevenson Services began establishing branch offices for sales, service and parts. R.N. "Dick" Conolly, who was a lieutenant colonel in the Air Force before joining the company in 1945, moved to Corpus Christi, Texas to be responsible for sales, service and parts for the South Texas territory from Victoria to the Rio Grand River. He was instrumental in setting up the first branch in the Rio Grande Valley at McAllen and the second in Corpus Christi.

In Corpus Christi, Lewis L. Bailey was named service manager for Diesel and gasoline engine maintenance,

By 1950, the number of Stewart & Stevenson Services employees had grown to include, among others: *Left to right:* Red Warren, Chauncey Beal, Bo Matthews, Jack Mahan, Joe "Pappy" Lawrence, unidentified, G. B. "Shan" Shandley, T. McArdle, Fred Mitchell, L. L. Bailey, Willie Hines, and Jack Lofton.

The first Stewart & Stevenson Services branch, established in McAllen, Texas in 1946.

The first Corpus Christi branch office, opened in 1947.

The Corpus Christi branch employees. Seated, left to right: Mickey Elliot, Bill McMenomy and "Val" Valentine. Standing, left to right: Mandy Leal, Harold Priesmeyer, E.L. "Easy" Elzner, N.N. Elkins, Roy Markwardt, and Bill Wiese.

and E.L. "Easy" Elzner joined the service department as mechanic. N.N. Elkins, hired to work in sales, sold Diesel engines to tugboat operators and shrimpers along the Texas Coast as well as Diesel power for cotton gins and oilfield drilling rigs. He became Corpus Christi branch manager when Conolly later took over responsibility for all branches and dealers.

Mechanic C.F. "Chauncey" Beal was transferred from the Houston Service Department to McAllen, where he was the service manager and also served as acting branch manager until J.C. "Jim" Brown was hired as branch manager and salesman.

Fred Malley, named sales director for the new Livingston branch, moved to Livingston to work the sawmill territory, working out of his house. Mike L. Ferguson was the maintenance, repairs and parts manager in Livingston.

Other branches were later opened in Dallas, Beaumont, San Antonio, Odessa, Lubbock, Wichita Falls and Amarillo.

Dealers who handled Stewart & Stevenson products were located in Laredo, San Antonio, Port Arthur and Longview. Others would soon follow.

A.B. "Dick" Grant, returning from four years in the U.S. Army, became the sales manager for the Rio Grande Valley in Laredo.

L.L. "Lee" Howell, a Diesel salesman, was named sales representative for the Wichita Falls office to serve northwest Texas.

Joe Manning and Dick Conolly raise the flag on the second Corpus Christi branch in 1951. Others, from left are: Hap Palmer, N.N. Elkins, "Pappy" Lawrence, G.B. Shandley, Charlie Stewart (from Detroit Diesel), C. Jim Stewart II, and two unidentified. Right of Manning are Charlie Bolduc and Augie Brecht from Detroit Diesel.

The Wichita Falls branch (above) was located in Army surplus barracks.

Left: The first Lubbock branch, established in 1950.

A growing Stewart & Stevenson staff attends the 1948 Sales Meeting. *Seated clockwise from left:* Lee Howel, R. N. "Dick" Conolly, N.N. Elkins, Dick Grant, Charlie Ames, E.E. "Slim" Childress, unidentified, Howard Perkins, Bob Mitchell, Jim Brown, C. Jim Stewart II, Floyd Smith, and Joe Manning. Standing at the podium is King Boyd. The other person standing is unidentified.

Engineering Department Formed

To facilitate expansion, Stewart & Stevenson Services established an Engineering Department to build on the ideas that had been developed during the war. The department initially consisted of E.E. "Slim" Childress, chief engineer, and Ernest McCombs, a draftsman.

Vertical Engine Developed

A new development was the vertical crankshaft Diesel engine for deep-well turbine pumps. Despite the fact that GM Detroit Diesel gave Stewart & Stevenson 46 reasons in writing why it couldn't be done, Fred Mitchell, who had a lot of practical experience with engines, assisted by Slim Childress, literally stood the Diesel engine on its end and modified it to drive a deep well turbine pump and for other applications. Because it did not have a right angle drive as on a conventional engine, the design simplified installation, increased efficiency and greatly reduced space requirements. It would later be applied to generator sets and fire pumps.

Engineered Power

"Engineered Power" — whether by Diesel, gasoline, butane or natural gas — became Stewart & Stevenson Services' trademark as the company set about to develop different types of engine-driven equipment. In addition to General Motors Detroit Diesel engines, which

One of the custom units Stewart & Stevenson built for the R. G. Wolff Studios in Hollywood, complete with diesel generator to provide on-location power.

ranged in size and models from 30 to 700 maximum horsepower, Stewart & Stevenson represented Continental Red Seal engines, Flagship Marine engines and Hallet Diesel engines, and later, Chrysler Industrial and Climax gas engines. The company also fabricated various types of electric switchgear, including switchboards and generator sets.

Government Contracts Continue

Stewart & Stevenson Services also continued seeking government contracts. General Manager Joe Manning, who had handled the government business during the war, turned that over to Charlie Ward, a War Production Board member hired by the company after the war. Ward remained in Washington until 1946, when he and Carsey Manning, Joe Manning's brother, opened the Dallas branch. Ward continued to handle the government business from Dallas. Carsey Manning, who worked briefly for the company in 1943 before joining the Navy and serving during the war, began his engine sales career in Dallas in 1946. He worked in sales in Tyler and Longview, covering the East Texas territory, then became the Dallas Branch manager, to be moved to Houston later, taking over responsibility for major contract sales and government contracts.

Custom Building Truck Bodies

Converting its giant remanufacturing plant on Harrisburg, where it had overhauled military trucks and jeeps, from war efforts to peace, Stewart & Stevenson turned the talents of the company's design engineers to fields allied with automotive body building. Facilities for the production and manufacture of custom built truck bodies were expanded.

Ross Stewart, President (1948 - 1956), Chairman of the Board (1956-1973).

Ross Stewart, First President

In 1948, Ross Stewart became the first president of Stewart & Stevenson Services. Well liked and respected by colleagues, city leaders and employees alike, "The Bear," as he was fondly called, would stride through the factory at Harrisburg wearing his famous straw hat, talk with the workers about their families and their jobs, and encourage them to suggest changes and improvements. His favorite question of employees was "What can we do to make things better?"

Devoted to each individual in the Stewart & Stevenson organization, which he referred to as the "Stewart & Stevenson Family," he was quick to criticize when someone was out of line or something was obviously wrong, but equally quick to praise and compliment when something worthwhile was accomplished. Under his leadership, the company continued to build on post-war innovations. By the end of the decade, the company had its first one million dollar revenue year.

Ross Stewart didn't limit his desire to make things better to Stewart & Stevenson only, but worked to make a better business climate for the nation as a whole. In 1947 he served as chairman of the U. S. Department of Commerce's small business advisory committee, which reported to Commerce Secretary Averill Harriman. The committee made a series of recommendations covering taxes, finance, business, management, trade practices, university extension services, government procurement and subcontracting, export and import problems, and publications relative to small business. The committee urged the Commerce Department to make the fullest possible use of chambers of commerce, trade and professional associations, libraries, business papers and other public interest groups.

Getting the Word Out

In the years immediately following the war, Stewart & Stevenson conducted an extensive campaign to "get the word out" about its products and services. Letters by post and wire, billboards, advertisements in newspapers, magazines and trade journals, brochures, and signs tacked to trees, barns and fence posts throughout Texas advertised Stewart & Stevenson. Every branch opening was featured in newspaper articles.

During 1948-1949, more than 2,000 Stewart & Stevenson signs were nailed to fence posts and tree trunks along highways throughout Texas to make the company name known throughout the state. A NOTICE accompanied the sign that read: "If this sign is objectionable, please tear it down. We had no intention to offend. Thanks." The vast majority of the signs remained for many years.

Advertisements in trade journals, such as *Rice Journal, Rice Grower, Gulf Coast Lumberman, Southern Fisherman, Oil and Gas Journal, Oil Weekly, Drilling, Diesel Progress, Diesel Power & Diesel Transportation, The Cotton, Cotton Oil Press, Slide Rule, Southern and Southwestern Bottler*, and newspapers throughout the state, generally included a testimonial from someone who had purchased a Diesel engine or power package for that particular industry. The ad also listed company locations in Houston, Dallas, Corpus Christi, McAllen, Wichita Falls, Greggton; its sales offices in Longview, Livingston, San Antonio, Beaumont and Laredo; adding "Distributors: for General Motors Diesel Engines, Continental Red Seal Engines, Flagship Marine Engines, Truck Bodies, etc.," and "Fabricators of: Electric Generator Sets, Electric Control Equipment, Portable Irrigation Units, Truck Bodies, etc." An appeal to others of that trade to contact Stewart & Stevenson would follow. Some examples:

Dale Washburn, president of the Cen-Val Gin of Donna, Texas said: *Stewart & Stevenson has the right answers to my power problems. Our gin has been powered with a Model 6-71 G.M. Diesel Engine which has given excellent performance and trouble-free service. Quite naturally when we considered enlarging our ginning facilities, we looked to Stewart & Stevenson Services Distributors for G.M. Diesels. They recommended a Twin 6-71 G.M. Diesel rated at 260 continuous horsepower. This G.M. Diesel provides an economical range of power which is available at the touch of a button. We could ask for no better service from our G.M. Diesel engine and the cooperation of Stewart & Stevenson Services.*

"Finding A Better Way" was more than just a slogan at Stewart & Stevenson. The company applied the concept to all aspects of its business and to all areas of industry and agriculture.

From the beginning, outstanding service provided by Stewart & Stevenson to its customers was a major contribution to the company's reputation and growth. At its 100th anniversary, service continues to be a vital part of the company's offerings. This photo is of an early "Field Service" truck on a customer's location in the early 1940s.

The ad for the truck body division in *Southwestern Bottler* read simply: *Let Stewart & Stevenson design and build the right truck bodies for you. Open or closed type bodies built to your specification. Rapid delivery. Stewart & Stevenson, 4516 Harrisburg, Houston, Texas —Anywhere – Parts – Service – Anytime.*

That slogan: "Anywhere – Parts – Service – Anytime®" was trumpeted on every occasion.

A 1948 advertisement picturing a service truck on the road in a stormy night featured the following copy: *Parts...Service, Day or Night, Rain or Shine — Our greatest concern is that you get parts and service for your G.M. Diesel when you want them, where you want them! Shutdowns on a location are expensive. So remember this vital service that's always available when you call STEWART & STEVENSON SERVICES.*

That advertisement would be repeated with an updated service truck many years later with the following copy:

When the Chips are Down and the Going's Rough — Engines are no better than the parts and service behind them. When a company can function as the world's leader in the Diesel engine and power business for over 30 consecutive years, there has to be some mighty good reasons, and we, the people at Stewart & Stevenson are well aware of them.

Our company has followed the practice, since the inception of our Engine and Power business, of rendering personal and individual service to our customers. Operating on the simple philosophy that "Engines are no better than the parts and service behind them," our company has become the world's largest distributor of Diesel engines and the foremost packager of Diesel engine-driven equipment in the world.

The reasons are simple: In the Diesel engine business, you don't use gimmicks. What you need is qualified people with the right attitude and a dedication toward customer service that knows no limits. Our policies at Stewart & Stevenson are such that our people take it as a personal responsibility to see that the customer gets a full measure for the dollar that he spends. The engine must be applied right for the specific application, it should be rated conservatively for its purpose, and it must be capable of delivering the needed, steady and reliable power suitable for the customer's application, and besides all this, we must assume the complete responsibility for service and parts regardless of where the engine may go.

Service to the customer has to be a way of life at Stewart & Stevenson. It's been that way for over 70 years — and we intend to keep it that way.

By the time the ad was repeated, Stewart & Stevenson had Diesel engine applications all over the world, and the company was using a world logo.

As Stewart & Stevenson expanded its Diesel engine line and applied it to an increasing number of industries, the company's advertising would reflect the progress.

Parts . . . Service
Day or Night
Rain or Shine
Our greatest concern is that you get parts and service for your G.M. Diesel engines when you want them, where you want them! Shut-downs on a location are expensive . . . so remember this vital service that's always available when you call STEWART & STEVENSON SERVICES, Houston.
DISTRIBUTORS
General Motors Diesel Engines . . . Continental Red Seal Engines
Hallett Diesel Engines
Anywhere SERVICE Anytime
G.M. DIESEL
RED SEAL CONTINENTAL
ENGINES
Stewart & Stevenson
SERVICES
HOUSTON - McALLEN - CORPUS CHRISTI - DALLAS

Fifty Years and Going Strong 1950 - 1960

From Ashes to World-Wide Power Distributor

On March 13, 1951, Stewart & Stevenson suffered a million-dollar loss when a fire destroyed a major portion of the Harrisburg plant, along with 70 new Diesel engines, 25 trucks, a number of large aircraft motors and several huge generators. With typical tenacity and vigor, Stewart & Stevenson employees, headed by Donald Stevenson, began cleanup before the smoke cleared. Sixty days later a new building was in place and manufacturing was restored.

A Golden Anniversary

Despite this setback, Stewart & Stevenson celebrated its 50th anniversary the following year as the world's largest distributor of Diesel engines. C. Jim Stewart & Stevenson co-founder Joe R. Stevenson blew out 50 candles on the anniversary cake, then he and company President Ross Stewart used the occasion to honor Charlie V. Foroi, the company's oldest employee, who had completed 32 years of service. The plaque presented to him was inscribed "in particular recognition of the major contribution you have made to our success."

Stewart & Stevenson also used the occasion to set forth a list of "Our Basic Principles":

1) Do unto others as you would have them do unto you
2) Select the right people for the right job
3) Provide training and education for our associates
4) Cooperation between departments
5) Encourage ideas, suggestions, criticisms and
6) Provide the means of assisting everyone in our organization to develop his security for himself and his family.

In its 50th year of operation, C. Jim Stewart & Stevenson and allied companies were controlled by Joe R. Stevenson, his son J. Donald Stevenson, and by second and third generation Stewarts, Ross Stewart and his son, C. Jim Stewart II.

Left to right: Ross Stewart, Joe R. Stevenson, J. Donald Stevenson, and C. Jim Stewart II celebrate the company's 50th anniversary.

Our Basic Principles

WE BELIEVE IN AND ARE PLEDGED TO:

1 Do Unto Others As Ye Would Have Them Do Unto You

There is no greater principle as a keystone to all other principles and implementations than our old Golden Rule. This philosophy must apply in all human relationships, whether it be with our own people, customers, suppliers, or the general public.

2 Select The Right People For The Right Job

Growth of an organization is dependent on care in selection of people who are physically, mentally and tempermentally suited for the position applied for, regardless of how big or how small the job. Select people who can develop and grow. Avoid square pegs for round holes.

3 Provide Training and Education For Our Associates

Provide every possible means for the training and education of everyone in the company in order that they may prepare and qualify for greater responsibilities as the opportunities occur.

4 Co-Operation Between Departments

Develop keen understanding between the various divisions and departments to the end that each can and will assist the other for the mutual benefit of all.

5 Encourage Ideas . . . Suggestions, Criticism

Provide facilities through suggestion box and other means for everyone to freely offer ideas, suggestions, or criticisms for improvement of our services, products and public relations, and for improvement of tools, plant and working conditions. Give due recognition for ideas.

6 Provide Means of Assisting Everyone In Our Organization To Develop His Security For Himself and His Family

Security, difficulty to define and impractical to give, rests within the hands of each individual. The company should provide means for the individual to grow and develop and thus create his own security through his own efforts. The company should provide:

(a) Opportunities for promotion and growth of each individual.

(b) Participation in profits due to individual efforts, wherever practicable.

(c) A method for participation in stock ownership.

(d) Attractive life and hospitalization insurance plans to assist on completing security programs.

(e) In short, it is the purpose of this organization, as far as security is concerned, to make available every opportunity possible for each individual to provide for his own security and not depend on others.

Our Basic Principles, established by company managers in 1952, are guidelines still followed by the company today.

1950s Goal: Diversify and Grow

During the 1950s, President Ross Stewart led the company in an all-out campaign to diversify and increase industrial and commercial volume, both at home and abroad. His goal was to develop new applications for new fields everywhere. To accomplish this, sales and engineering staffs were increased, and the number of employees approached 500. Salesmen traveled far and wide to seek out new markets for the products developed: to the new State of Alaska, on the European Continent, in the South Pacific and in South America. Among the products developed during the decade were sand-fracturing units, pumpers and blenders for oil well re-servicing operations, new applications of pumps and vertical engines for irrigation, airport ground power utility units of several variations, and refrigerated car units, all by the Engine Division of Stewart & Stevenson Services.

"My father did more than just aim the company in the direction he wanted it to go," C. Jim Stewart II recalled of Ross Stewart. "Dad was very aggressive, and he also had an appreciation for the abilities of the employees. His philosophy was that if an employee came up with an idea for a new product or service, he should be given an opportunity to develop it, and that philosophy paid off. He also believed in promoting employees when an opportunity came up. That's why our people have always been the most loyal, most valuable asset we have."

By 1952, a growing sales staff was traveling all over the world to sell Diesel engine applications. Attending the annual sales meeting in Houston are: *(seated, left to right)* Dick Grant, Dick Conolly, N.N. Elkins, Doug Chene, and a Chrysler Corporation representative. *Standing, left to right:* Unidentified, Fred Malley, Bob Hardy, unidentified, Emmet Dobbs, Harvey Trotter, unidentified, Bill Collet, Jim Brown, King Boyd, Bill Foster, N.J. "Tiny" Herman, Slim Childress, Tom Langham, Dick Morrison, Lee Howell, and Harold Speers.

The Price of Leadership

By 1954, 16 years after Stewart & Stevenson entered the Diesel engine business, the company already was the nation's largest distributor of General Motors Diesel engines. Engines sold by the company supplied power to drill for oil, gin cotton, power boats, build dams, irrigate farm lands, power generators and countless other installations.

Ross Stewart spoke at that year's annual sales meeting on "The Price of Leadership."

"What makes a leader?" he asked, then answered his own question.

"When I analyzed the reason for our leadership in the industrial engine field, the answer was very simple: in the beginning, we established an uncompromising set of principles...and we have followed them day after day for the 16 years we have been in the engine business. We can continue to hold our number one position indefinitely by continuing to follow these same principles with the same enthusiasm and sincerity that we have shown in the past...and continue to work harder than the competition."

The full version of his speech was published in a small booklet and given to all the salesmen.

Joe Manning and Ross Stewart visit with a beauty queen at the Detroit Diesel Engine Division booth at the GM Powerama in Chicago in 1955.

J. Donald Stevenson, President (1956-1973), Chairman of the Board (1973-1976).

1950s Milestones

Achieving the company's first $10 million revenue year topped Stewart & Stevenson's list of 1950s milestones.

The attainment of these revenues was aided by the formation of the Oilfield Equipment Division in 1954; the development of the Quad Series 110 Diesel-powered units in 1956; the formation of the Machinery Acceptance Corporation in 1957; entry into the aircraft ground support business with a revolutionary low-silhouette Diesel-powered ground power unit in 1958; and the addition of Hyster industrial forklift trucks in 1959. These developments would continue to fuel growth in decades to come.

By the end of 1959, Stewart & Stevenson Services, completing its first 20 years in the Diesel business, had sold and applied more than 15,000 Diesel and several thousand Chrysler industrial marine engines. Through 1959, in Diesels alone, the company had applied a whopping two million horsepower.

J. Donald Stevenson, President

In 1956, Ross Stewart became Chairman of the Board and J. Donald Stevenson took up the reins as President of Stewart & Stevenson Services. Under Donald Stevenson's leadership, the company made tremendous strides with innovations for the oil drilling industry, commercial airlines, commercial fishing, oil-well supply and crew boats, and backup electrical power systems. Many new products that came on the market in the late 1950s were greatly expanded upon in the 1960s.

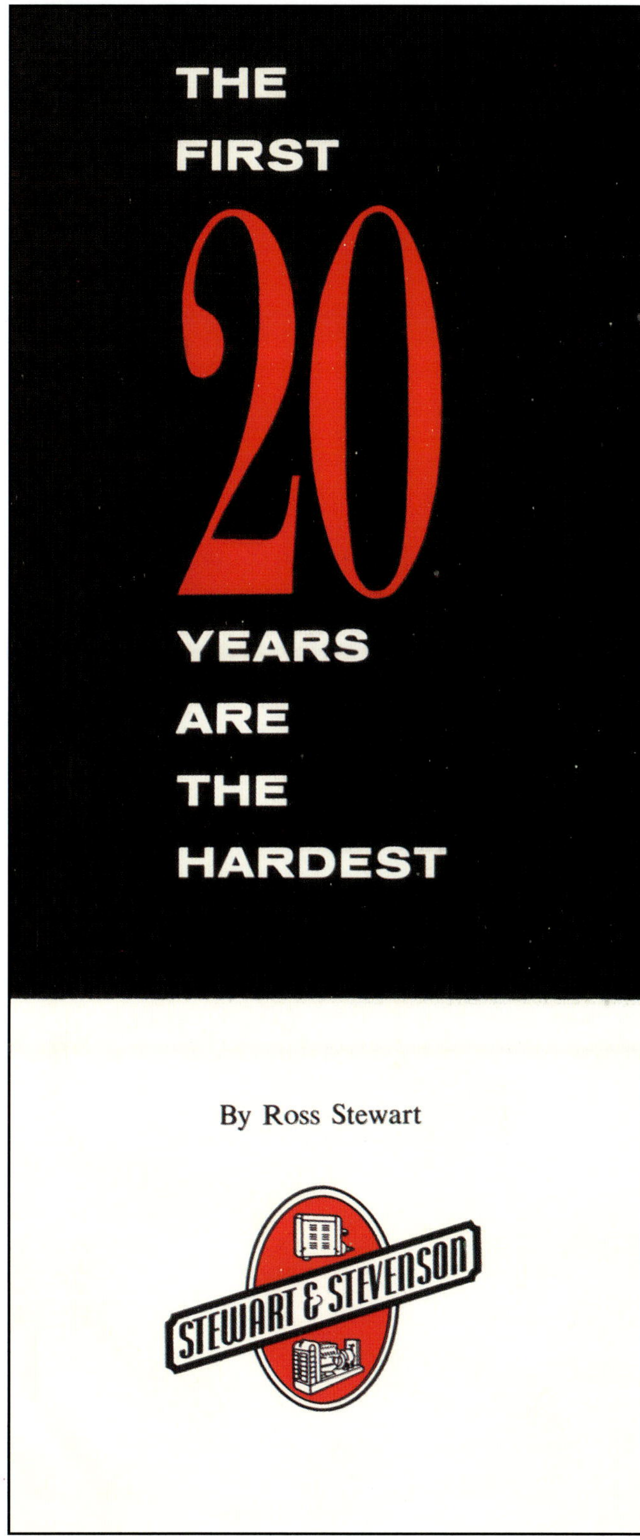

Above: The cover of Stewart & Stevenson Services' first annual report, 1956. *Right:* Ross Stewart's review of the company's first 20 years as a GM Diesel distributor, published in 1958, recounted the company's growth into the "World's Largest Distributor of Diesel Engines" and urged building on that success through "hard work, enthusiasm, aggressive sales approach, good advertising and publicity."

Machinery Acceptance Corporation Formed

To finance sales of the company's growing and now highly diversified product line for Stewart & Stevenson customers, in 1957 Ross Stewart formed the Machinery Acceptance Corporation. Patterned after the General Motors Acceptance Corporation (GMAC), whereby automobiles could be purchased on credit and paid out over a period of time, Stewart & Stevenson's MAC enabled buyers of its products to do the same.

Under the direction of L. O. Carlson, Machinery Acceptance Corporation established banking connections throughout all of Stewart & Stevenson's operating territories to provide the financing. The company's banking connections helped Stewart & Stevenson to implement credit policies by surveying the economic conditions connected with a particular industry in its territory.

Powering the Oil Field

By 1950, Houston, Texas was the center of the burgeoning petrochemical industry. Most drilling rigs initially had natural gas or steam engines and operated on free gas, the by-product of the oil or gas well. As Stewart & Stevenson Services salesmen successfully convinced drilling rig owners of the greater reliability and better overall performance of the Diesel engine, the company began providing the power required in the oil field. Stewart & Stevenson Diesel engines soon were providing auxiliary and prime power for drilling rigs all over the world.

Offshore drilling rigs, however, required a lot more power than conventional land rigs. In 1950 Stewart & Stevenson's largest engine, the Detroit Diesel Quad 71, consisted of four engines driving through a common gear box. It provided about 520 horsepower. If a drilling rig needed 2000 horsepower, Stewart & Stevenson would have to supply four Quad 71s, or 16 engines, which was actually too many engines to do the job.

The EMD Story

Stewart & Stevenson's competition had much larger engines. The Electro-Motive Division of General Motors built the 567 Series Engine, which went up to about 1600 horsepower. An extremely popular engine with the railroads, it powered 90 percent of all locomotives.

Deciding that the EMD 567 would provide the correct horsepower and make a fine engine for drilling rigs, Ross Stewart and Joe Manning traveled to EMD headquarters in LaGrange, Illinois, to see if Stewart & Stevenson could acquire the engine.

"The EMD people were proud people who weren't interested in anything but the railroad business. They didn't want to talk to us about selling us engines for industrial use, because they could sell all the engines they could make for locomotives. That was the business they knew," Manning said.

Persevering, Ross Stewart, who had gone to school

Two GENERAL MOTORS
Quad Single—6-71 Drilling Engines powering Deep Well Drilling Rig.
Location: New Mexico.
Operator: Loffland Co.
Well Depth: 14,000 feet.
H.P. (Each Quad)—520 continuous rating.
Weight: 12,440 pounds each.

An oil rig with a General Motors Detroit Diesel engine driven pump.

with the wife of Nelsen Dessendorf, EMD's general manager, tried to get him to discuss selling the engine to Stewart & Stevenson. Dessendorf refused Stewart and Manning an audience, instead referring them to his assistant. Stewart and Manning successfully sold the assistant on the idea that selling Stewart & Stevenson EMD engines for drilling rig power would be a great diversification for EMD. They assured him that Stewart & Stevenson would tailor the engine to fit the drilling rig, allaying his concern that EMD would have to custom make the engine for this application.

After several trips to Illinois, Stewart and Manning finally made the deal for Stewart & Stevenson to have the EMD engine for industrial applications worldwide, with the exception of marine and railroad applications.

Stewart & Stevenson did a lot of advertising and sold EMD engines for drilling rigs to Humble Oil & Refining Company, Magnolia Petroleum Company. Phillips and others.

Slim Childress, who sold six EMD engines, along with generators and motors to Humble Oil in 1957, said, "that $770,000 contract was the largest commercial order Stewart & Stevenson had ever had up to that time."

Stewart & Stevenson built the General Motors Detroit Diesel 6-110 into a Quad to make a 1000 horsepower package suitable for the offshore oil drilling business.

It was Stewart & Stevenson's efforts and persuasion that put EMD in the oilfield drilling engine market.

The 6-110 Quad

By this time, the General Motors Detroit Diesel Engine Division had a larger single engine, the 6-110, which was approximately 50 percent larger than the 71 series. Under the direction of Slim Childress, Stewart & Stevenson built the 6-110 into a Quad to make a 1000 horsepower package suitable for the offshore oil drilling business.

In the 1950s, C. G. Glasscock Drilling Company, owned by a successful Corpus Christi driller named Gus Glasscock, had built one of the largest offshore drilling rigs in the world, called Mr. Gus I. It was powered with Stewart & Stevenson Quad 71s, sold by N.N. Elkins out of the Corpus Christi branch.

"Glasscock had plans to build Mr. Gus II and wanted 5000 horsepower on it, which would be the most powerful drilling rig in the world," Manning said. "The Electro-Motive Division of General Motors and all of the oilfield supply companies were bidding on it, but we won the contract and provided all of the power for Mr. Gus II with our Quad 110."

Stewart & Stevenson soon dominated the offshore oilwell drilling market in the Gulf of Mexico. Of the 36 oil rigs Humble Oil had offshore, Stewart & Stevenson's Diesel engines powered all but two.

Oilfield Equipment Division Formed

"Shortly after World War II, when I got out of the service and went to work full time for the company," C. Jim Stewart II recalled, "Paul Koomey came to me and asked for a job. I had known Koomey in high school, and he also was a friend of Tom Langham. They had gone to Texas A&M together. After graduating from A&M, Koomey had gone to work on an offshore drilling rig, but his mother had become ill, and he needed a job in Houston so he could take care of her.

"I was surprised and flattered that he thought I had the authority to hire someone," C. Jim II said. "I didn't think I did, but I looked around. The manager of the Alemite Division said Paul could be the assistant shipping clerk. It was an all-day job, not skilled, but it was in Houston so he could take care of his mother."

Paul Koomey not only handled the shipping, but had time left over to help the service man repair Alemite pumps and equipment.

"He was very proficient, very quick in repair," C. Jim II recalled.

"About this time, the Stewart Warner people told our Alemite manager that Stewart & Stevenson's large territory required two salesmen. The Alemite salesman in Houston was always the number one Alemite salesman in the nation, primarily because he was the only industrial salesman on the Gulf Coast.

"We had promised Paul that if an opportunity came up, we would consider him for the position, so I told the Alemite manager that it was only right to let Paul be the new Alemite salesman," C. Jim II said.

The Alemite manager expressed some doubts about Koomey being the man for the job, to which C. Jim II replied, "Well, he's a damn good shipping clerk and repair man, so if he doesn't succeed as a salesman, we can always bring him back into the shop."

Ross Stewart supported his son's opinion. He told the Alemite manager to tell Koomey what he needed to do, give him support, and let him try.

"That's our policy," he said.

So the Alemite manager reluctantly agreed.

The Alemite territory was divided. George Darvy, the first Alemite salesman, gave Paul some of his accounts that he hadn't done very well with. One of those accounts was Brown & Root.

Soon after, one of Brown & Root's purchasing agents called Ross Stewart and said, "Mr. Koomey is a hell of a salesman. He comes in, tells me what I need, already has the order form filled out and I sign it. I don't even know what I'm ordering."

Two years later, Koomey was the number one Alemite salesman in the nation.

The Stewart & Stevenson Oilfield Equipment Division, formed in 1954, resulted from Paul Koomey's blowout preventer closing unit. The division became one of the company's most profitable profit centers.

Cameron Iron Works, which dominated in blowout preventers for drilling rigs, also was in Koomey's territory. He took an Alemite air motor that pumped fluids under high pressure to the Cameron Iron people and explained how they could use it to close their blowout preventers on drilling rigs. Their current blowout preventer had a large valve that had to be closed by hand. Few men wanted to hang around and close the valve when a well blew, so there was a critical need for a fast way to close the blowout preventer.

Koomey told the Cameron Iron people that this Alemite air motor would close the valve within three seconds of a blowout, but they wouldn't believe him. Anyway, they said, Cameron was working on its own industrial air motor to use on their blowout preventer.

"Paul came to us and explained what he had in mind regarding closing the valves, and said he wanted to build some in our truck body shop so he could show Cameron Iron how it worked," C. Jim II said. "We gave him a corner in which to build one, but before he could show it to Cameron, one of the oil rig owners bought it to use on a new rig. Every time he built one, a drilling contractor would buy it. So we wound up selling them directly to the contractors."

When Paul Koomey went to Ross Stewart and asked what he should call the new product, Stewart is said to have told him, "Your name is strange enough. Call it the Koomey unit!"

The Koomey blowout preventer closing unit, or the Koomey, as it came to be called, became the industry standard and today continues to be the generic name for all such units. It was the beginning of the Stewart & Stevenson Oilfield Equipment Division, formed in 1954, which became one of Stewart & Stevenson's most profitable profit centers of all time.

Oil Well Servicing Units

Stewart & Stevenson got into the production and sales of oil well servicing equipment after being contacted by Clyde Cecil and Dick Tunnel, who had left Halliburton to form O'FRAC, an oil well fracturing business. They asked Stewart & Stevenson to build two fracturing units for them, which the company delivered in 1953. Each unit consisted of an Oilwell Model 48PHD pump driven by a Twin 6-71 Detroit Diesel.

To get more oil from a formation, oil rig owners would pump a mixture of fluid and proppants — sand, gravel, or crushed nut shells — into the oil well at high pressure to "fracture" the formation, causing it to open like the pages of a book being fanned apart. The proppants would hold the fractures open, and as much as five to fifteen times the amount of oil would be pumped out of the formation.

On August 30, 1959, Stewart & Stevenson delivered the world's largest mobile sand fracturing blender to The Western Company of Fort Worth, one of the foremost oil well servicing companies in the Southwest. Designed and manufactured in its entirety by Stewart & Stevenson Services, the sand fracturing blender for The Western Company had twice the capacity of any other unit in the field. It produced 4,200 gallons of fluid per minute and handled over 16,000 pounds of sand per minute. Powered by a 280 horsepower turbo-charged General Motors 6-71 Diesel engine, the 35-foot long, 8-foot wide, 13-foot high unit had been tested for speed and maneuverability on Houston's freeways.

The sand fracturing blender for Western was followed by development of the world's largest oilfield pumping truck, called "Fieldmaster." It was followed by the even more powerful "Fracmaster" mobile pumper.

In 1961, Stewart & Stevenson received a $600,000 order from The Western Company for four new automatic sand blenders and five General Motors Diesel-powered double pumpers. Additional orders for oilfield fracturing and cementing units followed. In 1967, the company supplied Western with Fracmaster III-G pumping units with engines and related powertrain equipment — the most powerful pieces of Diesel-powered mobile pumping equipment.

Stewart & Stevenson oilwell servicing equipment grew to include trucks for logging, acidizing, cementing and blending operations, as well as fracturing. Today, the company is one of the largest manufacturers in the world for this type equipment.

Early oil well servicing units built by Stewart & Stevenson included sand blenders (top), cementers (above) and the Fracmaster III-G pumping unit (bottom).

The 'Cadillac' of Airline Ground Support

In 1952, the Hobart Company of Troy, Ohio, dominated the airline ground support electrical business. Hobart had garnered 98 percent of the business with its units, on trailers or self-propelled, that generated electricity for the aircraft's use while it was on the ground. Hobart's units were all gasoline engine powered.

Stewart & Stevenson decided to build a better unit using Diesel power. Diesel engines did not use a spark plug or carburetor, which frequently caused problems, and its fuel was cheaper. The Diesel engine also would run on the JP4 fuel used by the new jet powered aircraft engines that had just started flying, so the airline would need only one fuel on the tarmac.

Though totally unknown in the commercial airline business, Stewart & Stevenson had provided 700 aircraft ground support generator sets to the U.S. Air Force in 1947. Those units were fairly small gasoline/electric portable ground power units mounted on four-wheel trailers or dollies. They were designed to operate as ground starting equipment for aircraft engines and to furnish electric energy for testing landing gear, flaps, radio, radar and for charging batteries.

Stewart & Stevenson actively and aggressively entered the commercial airline ground support field in 1957 when the large jet liners replaced smaller piston engine aircraft and required a high-quality, precise-performance 400-cycle electric ground support unit. The design and manufacture of equipment for missile systems support gave the company vast experience in this 400-cycle generator field.

The Prototype

Led by R. W. "Bob" Mitchell, the Engineering Department built a prototype. It utilized a Model 4-71 GM Detroit Diesel engine, a high efficiency improved performance generator, an automatic transmission, and a front drive axle. This new, self-propelled package was rated at 90KVA, 400-cycle.

Built with a much lower silhouette than previous airline ground support units, the prototype was low enough to run under the wing of an aircraft, thus eliminating the chance for an accident that would damage an aircraft wing. Designed with fins and streamlined headlights like the expensive Cadillacs and Chryslers of the day, it also had more style than other airline ground support units.

Stewart & Stevenson began calling on the airlines: Braniff, Continental, Delta, Trans Canada, United, Air Canada, Eastern and others — but couldn't get to first base. So, Stewart & Stevenson flew the prototype to airline industry conventions and shows all over the country, displaying it outside where airline officials could see and drive it. With a stewardess or two at the wheel, the prototype began to draw a lot of attention.

One by one, Stewart & Stevenson convinced the airlines that its airline ground support unit was better equipment than the competition's and less costly to operate and maintain. Continental Airlines bought the first, Braniff International the second, then Air Canada and Delta. With these initial sales, Stewart & Stevenson went after the market in earnest.

Stewart & Stevenson initiated a development program that included a full line of Diesel electric generator sets for mounting on aircraft tow tractors, trucks, trailers and a wide range of sizes with premium performance features.

All had wheel wells to facilitate changing a flat tire. Ground power units for northern locations were equipped with cabs to protect the operator in cold weather, and those for Canadian airlines were designed with extra ground clearance to accommodate snowy conditions in Canada.

The Big Blue Machine (left) thrilled neighborhood youngsters when Tom Langham or C. Jim Stewart II drove it home for a weekend. *Above:* Art Jones examines a cab type ground power unit designed for cold regions.

The Big Blue Machine

(Told to Meridith Mendenhall, granddaughter to Tom Langham, by her mother, Susan Langham Mendenhall, and grandmother, Linda Langham Cole.)

When I was eight, and neighbors knew neighbors and the kids played outside in the front and in the street, my Dad, Tom Langham, surprised us all.

We lived on Norfolk Street, you know, where the Houston Rockets play. The Southwest Freeway was being built — Highway 59, to those who are new to town.

Our house was in the middle of the block, so it was a great meeting place for all the kids who lived on either side of us.

My sister, brother and I were playing in the middle of the street when we heard screaming at the end of the block. There was this huge rectangle, ugly blue thing, driving down the street. It did not look like a car, more like a moving box.

On the front left side there was a man with dark hair sitting in a little chair. It was my Dad.

Now my Dad was sort of large and this seat was really small.

Pretty funny looking, but the thing he was driving was even funnier looking.

He drove right into our driveway, with about 50 kids running after him.

He was driving one of the first Ground Power Units. All the kids in the neighborhood, including myself, thought he was the neatest, coolest Dad ever.

He took all the kids for a ride. I can imagine all the stories he told Bob Mitchell the next day.

What a treat for all of us!

Times have changed but every time I go to the airport, the first thing I do when waiting on a plane is go to the window and find that ugly rectangle thing and think of Dad.

Innovations Conquer New Airlines and Air Force

Air Canada came forth with a further diversification opportunity by adding a requirement for a 28-volt DC for certain piston engine aircraft, and 112-volt DC requirement for the Comet and Vanguard. Stewart & Stevenson responded by making a static-type transformer rectifier concept that soon was in universal use. Air Canada purchased 80 of the Stewart & Stevenson combination AC/DC vehicles to service their routes.

The AC/DC unit's increased weight resulted in some new problems in the vehicle drive line. The problem was exacerbated by the automatic shift transmission characteristics, which allowed personnel to "hot-rod" the self-propelled chassis. The Detroit Diesel's torque and acceleration was too much for the automatic transmission in the hands of inexperienced personnel. Stewart & Stevenson had to find a better way.

The answer came with the design and application of a hydrostatic (all hydraulic) drive in place of the automatic transmission. Now the airlines were able to operate the self-propelled units without the worry of burning out the clutches, stripping transmissions or differential gears. This simple drive application became the firm's standard mode for all self-propelled aircraft ground support equipment.

In 1964 Stewart & Stevenson designed and custom built for the U.S. Air Force a sophisticated aircraft ground service unit that performed five critical functions. The trailer mounted unit through separate systems: 1) drained and purged the aircraft with dry nitrogen, 2) vacuum pumped residual material, 3) replenished a special coolant system, 4) mixed and added a solution of distilled water and ethylene glycol for the

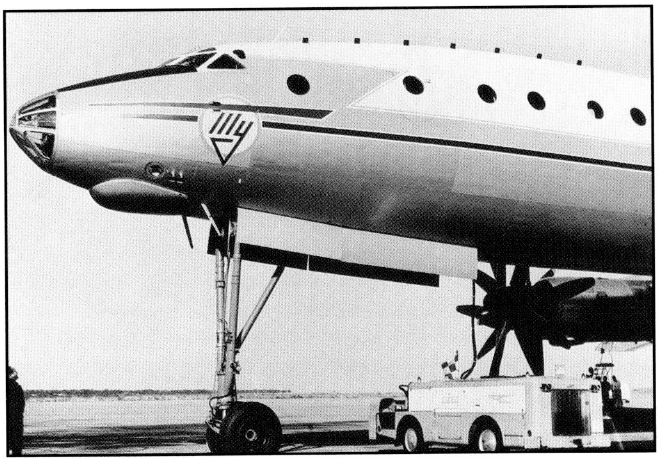

A Stewart & Stevenson ground power unit services a Russian-built aircraft.

system, and 5) provided a freon gas charge for the liquid and vapor cycles of the aircraft cooling system. The unit operated under temperatures that ranged from 20 to 125 degrees F.

Combination Units

In 1968, the first Diesel engine powered combination airstart and electric ground power units designed and built to serve the commercial airline industry's modern jet aircraft was introduced by Stewart & Stevenson. This unit was designed to provide high quality, constant uninterrupted electric power supply for jet aircraft while simultaneously providing high capacity compressed air for starting the jet engines. It was an "all-in-one" package. R. W. Mitchell, chief engineer, headed the project.

Keeping pace with the development of jumbo jets and supersonic transports (SSTs), Stewart & Stevenson introduced higher capacity, pneumatic ground power units for air conditioning service and for starting the 747 and SST. The new airstart unit was designed to provide a continuous air supply for the aircraft environmental equipment, cabin pressurization testing and operation of various aircraft systems as well as starting when required. It was available in two configurations: a dual arrangement consisting of two engine-driven compressor modules manifolded together and mounted on

Advertisement for the Combination GPU.

a single vehicle; and in a single, large engine-driven compressor unit. Each of the two-engine models delivered a continuous surge-free flow of air at 240 pounds per minute at 40 psig. The large single model delivered 270 pounds per minute at 41 psig.

With these new units, Stewart & Stevenson's airline customers grew to include TWA, Air West, American Airlines, Northwest Orient, North Central, Canadian Pacific, Alitalia Airlines and the Hawaiian Airline. Stewart & Stevenson gradually took the business away from Hobart and changed the whole aircraft ground electrical support industry from gasoline powered to Diesel-powered units. From this point on, Stewart & Stevenson and GM Detroit Diesel engines were the recognized leader in the aircraft ground support market.

Supporting U.S. Missile & Space Programs

After World War II, ideological differences between the U.S. and the Soviet Union and other Communist nations led to the Cold War that was marked by a massive military buildup of missiles and nuclear weapons on both sides, as well as the race to space.

In the early 1950s, Stewart & Stevenson began furnishing precise power generator sets that played some part on every one of the U.S. missile and space programs. The generator sets were manufactured at the Harrisburg plant, along with other power equipment supplied to the U.S. Army, Air Force and Navy for use in missile and radar work.

Joe Manning, vice president and general manager of Stewart & Stevenson at the time, was quoted in the February 1, 1958 *Houston Post* as saying that "highly technical" generator sets were specially designed, engineered and manufactured to lend ground support to missiles.

"Precise power — and not just any type of power — is a prime necessity in missile work," Manning told reporter Sam Weiner. He added that the firing of a missile involved "an entire series of events, many of which call for highly sensitive equipment."

The occasion of the article was the launch of the Jupiter-C satellite from Cape Canaveral, Florida, which was being tracked through space by equipment powered with intricate generator sets manufactured by Stewart & Stevenson.

The article noted that "in the launching of Atlas, Navajo and Redstone at Cape Canaveral, the Stewart & Stevenson units power missile control guidance systems, provide electricity for radar tracking devices and offer necessary checkout power while the missiles are being readied for launching."

Stewart & Stevenson also supplied power units for Redstone Missile Center in Huntsville, Alabama, on the West Coast, and at Ft. Bliss.

Operation Shade Tree

The U.S. Army Corps of Engineers in 1957 began purchasing generator sets for the U.S. Air Force Nike Missile Program. Stewart & Stevenson bid against Consolidated Diesel Electric Company, which was offering Cummins engine-powered units. Stewart & Stevenson got the first order for its Detroit Diesel powered units. On the second procurement, Condec won the order using Cummins power units.

The third time it came up for bid, Charlie Ward and Joe Manning came back with the contract.

"Slim, we have a problem fulfilling this contract," Slim Childress said Manning told him. "In order to have a lower cost, we bid using the 3-71 engine instead of the 4-71, but you can make it work, so get busy!"

Childress and the engineers built a generator set using the Model 3-71 Detroit Diesel as the prime mover and took it for environmental testing to Red River Pass, New Mexico. During the time he was at Red River Pass redesigning and re-engineering the Diesel generator, "Manning came up to see if we were ever going to finish and come home," Childress recalled.

"Joe took the cardboard out of a clean laundered shirt he had in his suitcase, wrote on the cardboard: OPERATION SHADE TREE, then nailed the sign on a tree," Childress laughed.

On October 8, 1957, Childress wrote on the sign "We did it!" then signed it and had Joe initial it.

After that, Childress said, the Army Corps of Engineers standardized on the S&S model and the company built over 4,000 of the generator sets for the military missile program.

The EMU-15, a 30kW generator set designed for the U.S. Air Force.

Emergency Power for Gemini/Apollo

In the early 1960s, GM Diesel engines were used extensively by Cape Canaveral and other space program installations to power key components in connection with the Houston-based NASA space probe work. GM Diesel units were used as ground power sources for intricate spacecraft launching, guidance and tracking systems and in protective warning networks. New lines of GM Diesel engines were introduced along with miniaturized control equipment for use in space probes work.

In 1964 Stewart & Stevenson completed an emergency power system for the Gemini and Apollo space probe programs at the Manned Spacecraft Center in Houston. The company engineered Diesel-powered generator sets for the Integrated Mission Control Center, which contained communications, simulations and training equipment for the two manned space programs. Uninterrupted power was of the utmost importance in the tracking of manned moon shots.

Charles L. Ward, vice president and general manager of Stewart & Stevenson Services at that time, said the units would serve as the center's total standby power system and would be completely automatic. Should there be any failure of the commercial power service, the generator sets would instantly start working to ensure a continuous supply at all times. Two Diesel units operated at half load so that if one failed, the other could furnish all required power without interruption. A third unit was a standby. A duplex panel to control all circuitry for the power supply was built in connection with the generator units.

The generator system powered tracking and computer devices that helped to guide and command the missions in space.

A minor power failure did occur during the Apollo 7 flight, but "in no way affected the mission," according to NASA engineers. A 90-second outage caused by a grounded fan motor in a cooling tower occurred near the beginning of the 11-day Apollo 7 mission. The power failure was automatically sensed and the emergency power system fulfilled the normal power requirements to the affected areas.

Housed in an 11,000 square-foot building adjacent to the Mission Control Building, the emergency power system was manned by a team of 35 to 40 persons who kept 24-hour watch during all spaceflights.

In 1968 Stewart & Stevenson provided NASA generator units for space programs at Cape Kennedy. These generator units, constructed for remote operation, were used as auxiliary power sources in connection with the Apollo moon exploration program.

Going International 1960 - 1970

Expanding Abroad and at Home

Stewart & Stevenson's annual sales meeting in February 1960 opened with greetings from "The Bear," Chairman of the Board Ross Stewart, who was in Bangkok, Thailand inspecting one of the company's power applications. The installation, under construction by Brown & Root, Inc., consisted of a battery of seven 1000kW General Motors Diesel generator sets that were providing power for the Yanhee River Dam Project.

Stewart & Stevenson products and services were being extended farther and farther afield, and around the world.

A 180-foot barge, with pumps and anchor winches Diesel-powered by Stewart & Stevenson, made its debut on Lake Maracaibo, Venezuela in 1960.

In Africa, a Stewart & Stevenson RIGELECTRIC engine electric power unit was supplying power for drilling operations on a Laughlin Porter oil rig.

In the Arctic, a 30kW Diesel generator set provided electric power around the clock to operate electronic apparatus vital to human safety and communications.

Four new Model 3-71N GM Detroit Diesel-powered units, designed and fabricated by Stewart & Stevenson, joined in Olympic Geophysical Company's search for oil and gas in the cold regions of Alaska.

Stewart & Stevenson International

To reinforce Stewart & Stevenson Services' sales position abroad and to go after a larger share of Western Europe's power equipment market, the company formed a new subsidiary, Stewart & Stevenson International.

Around Town

At home, Stewart & Stevenson delivered a specially designed emergency engine generator set to Houston's new Intercontinental Airport to automatically supply electric power in case of normal power failure. Prime mover for the 250kW generator, designed to operate completely unattended, was a Model 12V-71 General Motors Detroit Diesel engine. An earlier model generator set supplied by Stewart & Stevenson to Hobby Airport was still in operation after 13 years. By far, the majority of the hospitals in the Texas Medical Center also relied on Stewart & Stevenson for standby power, as did NASA, whose communications with spacecraft carrying astronauts into space demanded 100 percent precise power.

Construction of Houston's Astrodome, ongoing in 1963, also was being aided by Stewart & Stevenson equipment. The company's caisson drilling unit with a rolling quill, utilized to drill all the footings for the domed stadium, saved Weber Drilling Company 30 days of contract time. The exclusive Stewart & Stevenson design, along with a General Motors 4-53 Diesel driving through a CRT-3331-3 Allison full power shift transmission, helped change the concept of foundation and footing drilling.

Stamped on each piece of equipment, whether shipped far away or used locally, the name Stewart & Stevenson Services provided the customer assurance of sound engineering, and complete facilities, parts and service responsibility.

Stewart & Stevenson's RIGELECTRIC powered drilling rigs around the globe.

Emergency generator sets for Gemini and Apollo were installed at NASA's Manned Spacecraft Center in Houston.

'Great Balls of Fire'

One stormy night in early September 1961 — we were in the midst of Hurricane Carla — I was just finishing my evening meal when the phone rang. It was Owen (Owen Metcalf, plant superintendent for more than 42 years) asking me to come back to the shop to help load a Quad Generator set for the Dow Chemical Co. near Freeport.

By the time I arrived, the crew, consisting of Owen, Paul Jennings, Howard Gordon, a truck driver and his swamper, were grouped around the generator set. It was lined up under the Craneway, which was an extension of Felix Street.

Owen decided I was to climb up the ladder and run the crane up to the front end of the 25-foot-long, 30,000-plus pound generator set. They would then secure a cable to the box base and I was to lift up the front end and hold steady while the trucker backed under a few feet so he and his swamper could latch on with the winch line. Then I was to let off and move the crane to the other end of the unit and lift up until the unit was level so they could suck the set to the center of the trailer.

I hadn't said a word, but I was looking at the high-voltage power lines coming into the plant from Felix Street. The lines were swaying back and forth in the 75 to 80 mph wind. Tin was rattling all around us, the lights would flicker, and I could see myself standing on the small platform allotted the crane operator if the power failed, stranding me there the remainder of the night and however long it took HL&P to fix the power lines. However, my personal problem was solved when a gust of wind blew the power lines together, thus causing a couple of fire balls 10-12 feet in diameter, and ending all power to that end of the plant.

We drove our cars into the Craneway area, parking them so the head lamps were facing the unit from both sides. Then I got on another Hyster and helped Howard Gordon lift the unit, one end at a time, and with the tractor winch line, we managed to load the unit.

I heard later that the trucker got to Dow's plant three days later. The unit was never placed into service.

One bad day and night at S&SS on Harrisburg Boulevard.

(By: John C. Metcalf, brother to Owen, retired after 42 years of service.)

Stewart & Stevenson generators supply emergency and backup power for airports, hospitals, business and industry during brownouts and Gulf Coast storms such as Hurricane Carla in 1961 and Hurricane Celia in 1970.

Harrisburg Blvd. Expansion

In 1965-66, C. Jim Stewart II, vice president, began overseeing a three-phase expansion program to enlarge and modernize the offices and plant facility of Stewart & Stevenson Services, Inc. at 4516 Harrisburg Boulevard.

The first stage added 16,500 square feet for assembly of government and major contract generator sets. The second stage was a 24,000-square-foot building to house the Diesel rebuilding center and truck engine replacement facility. The third phase, a 17,000-square-foot structure, provided space for complete new engineering facilities as well as factory production offices, the purchasing department and parts accounting.

The additional 57,500 square feet were designed to help the company adhere to stricter quality control procedures, better utilize manpower and move toward better coordination of production and scheduling.

Congress/Preston Ave. Expansion

In 1965, Stewart & Stevenson completed a $1.5 million expansion and remodel of the entire 1700 block of Congress and Preston Avenues where the original partnership of C. Jim Stewart and Joe R. Stevenson began in 1902. The business that began in a blacksmith shop was now an engine power and transportation enterprise with 10 branches around the country and worldwide annual sales exceeding $30 million. The full block, bounded by Preston, Chenevert, Congress and Jackson, was occupied by a two-story complex of sales, service and administrative facilities. Housed there:

1) Alemite Company of the Gulf Coast, since 1925

This ad ran in the Sunday, August 22, 1965 *Houston Chronicle*.

the exclusive distributors of Alemite and Stewart-Warner products along the Texas Gulf Coast, Louisiana and Southern Mississippi.

2) C. Jim Stewart & Stevenson, Inc. Truck Lease Division, the largest and most modern truck rental, leasing and repair facility in the Southwest.

3) C. Jim Stewart & Stevenson Southwest Company, Materials Handling Division, with a full line of new, re-manufactured, rebuilt or used lift trucks. The division offered complete sales and service of gasoline powered Hyster lift trucks and electric powered Raymond fork trucks.

4) Stewart & Stevenson Distributing Company, Paint and Automotive Division, distributor of DuPont automotive paint for more than 40 years and the leading distributor in the Southwest. The division was also a distributor of a variety of shop tools and accessories.

5) Stewart & Stevenson Distributing Company Oilfield Specialty Division, which sold its products through oilfield supply stores.

Government, Major Contracts Relocated to Houston

In July 1969, Stewart & Stevenson's Government and Major Contracts Sales Office, headed by Carsey Manning, was relocated from Dallas to Houston to allow a closer working relationship between that office and the Houston factory and administration.

1960s Innovations

To meet the rugged requirements of railroad service, Stewart & Stevenson Services in 1961 developed the Under-Car Generator Set. This application allowed conversion of ice bunker cars to Diesel-electric powered refrigeration types without loss of cargo space at either end of the car. Prime power was furnished by a General Motors 2-cylinder Diesel engine. The refrigeration sys-

By 1962, more than 1000 Stewart & Stevenson Diesel-electric under-car generator sets were refrigerating rail cars.

Stewart & Stevenson entered the gas turbine-powered equipment market in 1963 with the manufacture of an oilfield cementer powered by gas turbines.

tem was mounted in the car wall to allow maintenance from the outside and the undercar Diesel generator set could be serviced at the ground level. The Stewart & Stevenson design met the clearance requirements of the American Railroad Association. More than 1000 units were in operation in 1962.

A major step that led to new product lines—and later a new division—came in 1963, when Stewart & Stevenson entered the gas turbine-powered equipment market with the manufacture of an oilfield cementer powered by gas turbines. Two 300 horsepower Boeing turbine engines drove two quintiplex pumps through a five-speed Spicer mechanical transmission. The engines weighed 350 pounds each and each transmission weighed 537 pounds. A 6V-71 Detroit Diesel under the cab drove all the ancillary equipment hydraulically.

That same year, the company introduced Showermaster, a semi-automatic wheel-move irrigation system. Stationary when sprinkling, it propelled itself to the next setting on lugged wheels. An end drive power control unit permitted the operator to control the system from the end of the line nearest the water supply valve.

In 1966, lightweight Diesels for small pleasure boats were introduced into the company's established line of marine engines under the registered trademark name of DIESELDRIVE®.

In 1967, Stewart & Stevenson Oilfield Division conducted sea trials of an Atomic Powered Wellhead. The acoustical controlled nuclear powered subsea wellhead, 90 feet deep in the Eugene Island area off the coast of Louisiana, opened a new era for the Oilfield Division of Stewart & Stevenson Distributing Company in the field of nuclear powered subsea controls. The equipment worked in depths up to 30,000 feet, with depth limited only to the packaging of the equipment.

Training the Family for the Future

Stewart & Stevenson's rapid growth in the 1950s and 1960s increased the need for well-trained employees. As a family company, Stewart & Stevenson looked to its family of employees as a source for future labor.

In 1962, the company inaugurated a summer vacation training program for the high school and college aged sons and daughters of company employees. The young women were provided general clerical jobs and the young men performed mechanical and electrical assembly, sheet metal work, warehouse control, painting and drafting. In 1968, 98 students participated in the summer training program.

Over the years the program, still in effect today, has provided Stewart & Stevenson with a large number of high caliber, loyal employees who got their start with the company this way.

Jimmy, Rick & David Stewart

At ages 10, 9 and 8, Jimmy, Rick and David, sons of C. Jim Stewart II, had their first experience working at the company. They worked one five-day week and were paid in cash, 50 cents an hour, for a total of $20 each for the week.

"We were given an old gasoline truck to sand off the paint so it could be repainted," Jimmy (C. Jim III) recalled, adding that they were all ready to quit long before the week was over. But when they quickly finished the first truck, they were given another project — to sand the box trailer on the back end of another truck. To the consternation of George Felts, their "manager," the ambitious boys sanded the whole truck, whose front had just been repainted and stenciled with the owner's name.

"Each year as we grew older," Jimmy recalled, an additional week was added to our 'employment contract' until we finally were working all summer."

Rick (Richard Ross) Stewart said that his father and grandfather would tell someone like Owen Metcalf "not to hesitate to whip us if we got out of line."

He recalled separating a lot of nuts and bolts, but said his best memories are of "the feeling of pride it gave me to drive to work with my father each morning and work with people who treated me just like one of the other guys working.

"It was part of the early gaining of our love for the company and, though perhaps not intentional on my father's part, making us decide that working for the

Early participants in the Stewart & Stevenson summer training program.

Jimmy, Rick and David Stewart get their first taste of working for the company during the summer of 1958.

company was what we wanted to do," Rick added. During his high school years, however, Rick decided he wanted to make more money than his father offered, so he went to work for Bellows Construction Company, which was building downtown skyscrapers.

"In those years," Rick explained, "my grandfather and the company were synonymous. If you thought of Stewart & Stevenson, you thought of the Bear, and if you thought of the Bear, you thought of the company. He and I loved each other, but I was the black sheep in the family. I had long hair during my college years, and he gave me hell about that. But actually, he liked it that I didn't just roll over and do what he wanted."

David Rial Stewart remembers "sweeping that downtown block (Preston and Congress) a thousand times, painting trash cans and doing other odd jobs" before migrating to the Harrisburg plant where, as he got older, he worked in the parts and shipping department pulling parts and shipping them to dealers and other branches.

"The 50 cents an hour we earned was a lot of money for an eight-year-old kid when a movie cost 25 cents and a Coke was 5 cents. A kid could have a really good time on 50 cents in the late 1950s.

"Grandfather (Ross Stewart) encouraged us to work summers and save our money. He said that for every share of Stewart & Stevenson stock we bought, he would match it. I didn't have anything better to do with the money, so I followed his advice and bought Stewart & Stevenson certificates. At the time, it was just a piece of paper, of little value. When the company went public and had a split of the original shares, the number of shares that I had earned and my grandfather had matched turned into a relatively large number for the amount of work that I did. That was one of the best investments I ever made," David said.

Garth Bates

Garth Bates, another grandson of Ross Stewart, remembers his first experience working at Stewart & Stevenson in the summer of 1959.

"The company had bought all the Detroit Diesel engines out of the Greyhound bus fleet and was converting them from bus service to industrial service and parts. We disassembled every engine into its most elemental parts, and the Diesel engine had many!

"Jimmy and I would get so dirty, the local sandwich shops — which, may I add, weren't much — wouldn't let us come in to buy lunch, and C. Jim II, the Chairman, who chauffeured us to work and back home, carried cardboard in his trunk to put on the floorboard and seat so we wouldn't track up his car. When we got home, our mothers wouldn't let us into the house with our clothes on.

"My beginning memory of work was that, if this was what it was all about, I definitely wanted to get an education because this wasn't something I wanted to do for a long, long time. But it was very instructive," Garth said, adding that his grandfather tended to agree with his assessment of his abilities.

"Son, you had better find a job that doesn't require the use of hand tools," Bates recalled his grandfather commenting as he watched him and Jimmy work on those engines.

Don and Keith Stevenson

Donald Edward and Keith Thomas Stevenson, sons of J. Donald Stevenson, also spent their high school and college summers working at Stewart & Stevenson.

"Every summer while I was in high school, I was 'vice president' in charge of moving 140-pound flywheels in the 140-degree heat at the Harrisburg plant," Keith Stevenson joked. "At the end of the day, I would be covered in grease from head to toe."

One special project assigned to them in the early to mid-1960s reads like a Tom Sawyer/Huckleberry Finn adventure. That job was to deliver 42-foot crew boats built by Breaux Baycraft at Loreauville, Louisiana on Bayou Teche to Houston, Galveston or Corpus Christi to show to a customer.

"These were brand new, bare boats, with no radios or telephones, no radar or navigation instruments," Don Stevenson recalled, "equipped only with Detroit Diesel engines. It took about 24 hours to make the trip on the Intracoastal Canal, and there was no place to stop and rest, so we usually just ran straight through."

Keith Stevenson picked up the story. "On one of our trips, going through the Vermilion and Calcasieu locks just east of the Calcasieu River on the Intracoastal Canal, the lock tender asked for a form that showed ownership of the boat. It was a brand new boat, and we obviously did not look like the usual oilfield hands. He had never seen a boat like this before and wanted to know all about it. When we told him it belonged to Stewart & Stevenson, he scratched his head, trying to remember what company that was. Finally, he said, 'Oh, I know Stewart & Stevenson. That's the outfit that owns General Motors!'"

Third Generation Stevenson Joins The Company

Donald Edward Stevenson, oldest son of J. Donald Stevenson, joined the company in July 1967 after graduating from Lamar Tech University with a degree in electrical engineering. He began his career in the Engineering Department where he helped design and build numerous projects, among them the bowthruster for Exxon and remote controls for oilfield frac units. From 1973 to 1984, he served as manager of the Electronics Department, which designed and built remote controls for power generation equipment, including the gas turbines on the Navy Aegis class ships. In 1984 he was named Vice President for Overall Quality Control, a position he still holds. Don Stevenson also serves on the Stewart & Stevenson Board of Directors.

Stevensons' Early Career Stories

One of Don Stevenson's early engineering projects was the design of the bowthruster.

"The vertical engine bowthruster concept was Bob Mitchell's, and B. J. Dodds was the mechanical engineer on the project," Stevenson said.

Keith Stevenson Joins Company

Keith Thomas Stevenson, younger son of J. Donald Stevenson and third generation family member, joined the company in June 1969 after graduating with honors from San Jacinto College and Sam Houston State University, earning a B.B.A. He began his career with the company in marine sales. In 1980 he became General Parts Manager of the Houston Branch, then moved to Dallas in 1983 to serve as assistant to the Vice President of Dealers and Branches. He became Vice President of Dealers and Branches in 1985. Since 1985, he has served as a vice president with continually changing responsibilities. He currently oversees General Parts and Service, serves as the company's liaison with the Hyster Company, and is responsible for the company's Technical Training Center in Houston.

As the electrical engineer on the project, Don was called upon to troubleshoot some bowthruster units on a Mississippi River tug that was pushing a 1200-foot-long string of Exxon barges carrying 7.5 million gallons of gasoline. He rode the barge from Baton Rouge to various points along the Mississippi and Ohio Rivers, and as far north as Boomer, West Virginia on the Kanawha River. This was the kind of service Stewart & Stevenson customers expected and for which the company was famous.

"One of my most memorable engine sales was to a guy building a shrimp boat," Keith Stevenson recalled. "Four months after we delivered the engine, he still hadn't paid for it, and John Kelly, our credit manager, told me I had to bring him the engine, the boat or the money, and he wanted it within a month.

A fuel barge, like the ones towed by the tug Don Stevenson was on.

"Out in the plant one December day, I moaned about my predicament, wondering what I was going to do. An engine cost a lot of money and I couldn't pay for it. One of the guys in the plant said his son worked for a local towing company and, if I wanted to repossess the boat, all I needed to do was call the towing company and have them tow it wherever I wanted it. I couldn't believe repossessing a boat could be that simple, but I had to try it because I was out of options."

Figuring the owner wouldn't be anywhere near the boat on Super Bowl Sunday, Keith decided that would be a good time to attempt repossessing it. He called the towing company and asked them to tow it from the mouth of the San Jacinto River where it was located to Mangone Shipbuilding. They agreed to do so as soon as the game was over.

With the winds gusting up to 35 mph, Keith waited for them in a utility boat in the middle of the ship channel in front of Mangone Shipbuilding.

"The deckhand on the tug threw the rope to me, didn't ask my name or anything. I asked him if he needed to see some identification, or did I need to sign anything. He hollered at me, 'Do you want this boat or not?' I said, Yes!, and he took off. With the wind gusting, it took me two hours with that little utility boat to tow the shrimper's boat 200 yards into the shipyard and get it tied up. The next month I got a bill for $75 from the towing company.

"That little repo job made me and my credit manager very happy. It also made the bank happy, because the shrimper also owed the bank for the loan he had used to purchase the boat.

"Everybody was happy but the shrimper," Keith concluded.

Power Spreads the News

The first issue of *Power*, Stewart & Stevenson's quarterly publication, appeared February 26, 1960. Its two-fold purpose was to keep employees of the company informed of company achievements and policies, and to keep friends and customers abreast of Stewart & Stevenson's progress in the area of power applications.

"In short," its editors proclaimed, "this will be a record of achievement."

Published through 1997, *Power News*, which it was later renamed, provided nearly four decades of Stewart & Stevenson's history in the power industry.

Competitive Spirit Increases Sales

Stewart & Stevenson's competitive spirit in all its fields of endeavor were personified by Chairman of the Board Ross Stewart's final remark at the 1968 Engine Division Annual Sales Meeting:

"Remember our share of business is all of it."

Ross Stewart frequently reminded salesmen of the goal: "Our fair industry share is 100 percent."

Taken to heart by company personnel, Stewart & Stevenson accomplished its first $50 million revenue year during the 1960s.

Ross Stewart: No Excuses Accepted

Just as Ross Stewart expected 100 percent effort from company salesmen, he also expected 100 percent effort from all other areas of the company, and had no tolerance for excuses, as this 1967 memo to Joe Manning illustrates:

Joe, there are a couple of subjects about which I want to go on record:

Number One, in my opinion, we are getting organized to the point we can't get anything done out there. I give you two outstanding cases:

One is the U.S. Engineers job which has been tied up for four weeks due to a circuit breaker problem which in my opinion is two-thirds imaginary and the other third could be cured with a certain amount of common sense which we have done before and built a reputation by so doing. But the fact we have an organization that will permit a contract to be tied up that is that important for such a length of time is a reflection on everybody from the top to the bottom. It is happening only because we allow it to happen. Certainly we are going to have trouble, but those things must be resolved in a practical, sensible way quickly and those that can't do it, must give way to those who can.

Now this is a criticism and I think a justifiable one, but to answer this, my suggestion is to get back to more practical people to handle our test line, our quality control or inspection or whatever you call it, and have somebody out there who has the courage to say what must be done and when it must be done and do not accept defeat or failure. This is a weakness we have never put up with and we can't afford to now. Now I am certain I have heard all the reasons why this has been delayed and permitted. None of these reasons make sense and you know it and I know it.

Now the next problem is that Navy job which has been in there for over a year. That has been started and stopped, pushed forward and pushed backward and delayed and horsed around until everything in it is completely out of line from the standpoint of getting something done. To cap the climax, they built one that would run, and then brought the second one up that wouldn't run — it wasn't wired properly. If they can build one right, why in the hell can't they build another one the same way? You are getting engineering excuses and you can't stand that. Now they are going into a "bugging project" on this unit. They don't need any bugging project because if you have one built right, then you should check to see why it was built that way. You will have some changes because of leaks and some minor changes, but they should know we are wrong and we don't want that.

We have been through this before and I don't think we can afford these type problems. Someone says we are doing better work than before through this quality control and a few other things. I will say if you can do it as well as you did in the past when you were building your reputation and could get the work out, then you are doing it better and if anything you should have improved products and improved quality. We cannot afford to stop production and anyone who gets in the way of that is the wrong person in the wrong place.

If you are interested in any further evidence of what I am talking about or if you don't understand, I will be glad to come out and give you a straight forward honest example of what I am talking about. I don't have to convince you, you know better than I what I am talking about. We cannot now and we never could tolerate this type operation. It appears you need just one person there and only one who takes a position and can't be altered at all by anybody and that is a schedule when every unit we are building is moving out and if it doesn't, then I mean someone is in real trouble.

New Agricultural Application

Showermaster Introduced

In 1965, Stewart & Stevenson introduced the "Showermaster," the company's uniquely designed semi-automatic wheel move irrigation system. Touted as the "ultimate answer to irrigation on America's labor-short farms," it was stationary when sprinkling, then would propel itself to the next setting on lugged wheels. An end drive power control unit permitted the operator to control the system from the end of the line nearest the water supply valve.

Hundreds of these systems, conceived by Bill Collett, Stewart & Stevenson's irrigation sales manager, were sold throughout the West and south central U.S. and accounted for a sizable amount of business.

In 1966 R. N. Connolly, vice president of the branch office in Corpus Christi, cited three Stewart & Stevenson salesmen for top sales in irrigation equipment: Joe Ray and Michael L. Molak of the Odessa General Irrigation Division, and Harold Priesmeyer, truck and irrigation sales engineer at Corpus Christi.

Showermaster self-propelled itself from one end of a field to the other on lugged wheels, irrigating the crop along the way. Note that the wheels are similar to the buggy wheels which the company's founders once manufactured.

Serving the Marine Industry

Packaged Boat Plan

Devised in 1958, the Packaged Boat Plan was just another of Stewart & Stevenson's innovations designed to sell more Diesel engines and thereby enhance the company's profits. Contracting for the purchase of a hull, which took six months or more to build, and furnishing the propulsion and auxiliary generator sets for installation in the hull, the company eliminated competition from other engine companies. Also, by having the hull well under construction prior to selling it to a purchaser, the time frame from receipt of the order until delivery to the customer was shortened. This was a big plus in selling the vessel, since purchasers of these large offshore supply boats often waited until the last minute to issue a purchase order —and then wanted delivery as soon as possible.

In 1958, under the direction of Ray Loden, Marine Sales Manager, Stewart & Stevenson christened the *Donald Stevenson*, one of the first of the company's offshore supply vessels built under the Packaged Boat Plan. Stewart & Stevenson purchased boats from various shipbuilders, including its own subsidiary, Mangone Shipyard, powered them with Diesel engines and sold or leased them in a packaged deal whereby the company provided single responsibility for the entire vessel.

Indestructible

In cooperation with KN Dynamics Company, Stewart & Stevenson designed a new extra heavy duty marine gear for use with GM Detroit Diesel Model 12V-71 and 16V-71 propulsion engines. Manufactured by Stewart & Stevenson, it was to be marketed under the trade name "Indestructible" and unconditionally warranted for one full year. In 1962 the tug *Matagorda* was the first to be fitted with the new Indestructible marine gear. The *Matagorda's* main engines were a pair of 16V-71 Diesels, each of which developed continuous 457 horsepower.

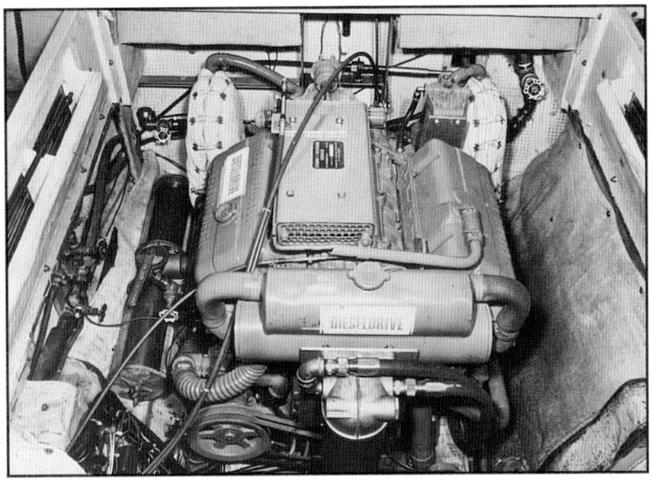

The inboard portion of DIESELDRIVE, a customized Series 53 Diesel engine for small pleasure crafts.

DIESELDRIVE

In 1966 Stewart & Stevenson pioneered and introduced lightweight Diesel marine engines for small boats, 20 feet and longer. These engines were marketed under the trade name DIESELDRIVE®, which was a registered trademark of the company.

One of the earliest offshore supply vessels built by Stewart & Stevenson under the "Packaged Boat Plan." The owner christened it *Donald Stevenson* in honor of J. Donald Stevenson, then company president.

DIESELDRIVE was offered in two sizes, with the Model 3-53 GM Detroit Diesel engine or with Model 4-53 engine. The 3-53 turned the same size propeller as a 155 horsepower gasoline engine. The 4-53 was equivalent to 200 horsepower gasoline engine. Each operated on approximately one-half the fuel used by the gasoline engine to do the same job. Diesel, a lower volatile fuel, also was much safer to use.

DIESELDRIVE units were designed for pleasure boats in the 20- to 30-foot class and houseboats of larger size. Supply boats and crew boats built under Stewart & Stevenson's "Packaged Boat Plan" were in use just about everywhere, in inland waters, offshore in the Gulf of Mexico, and from Alaska to Puerto Rico, to Norway and even Japan. Safe, dependable, and economical, they paid off for owners of boats of many types.

A pleasure craft powered by DIESELDRIVE.

The *Jupiter*, a 152-foot twin-screw Diesel motor utility and supply vessel built by Stewart & Stevenson's subsidiary Mangone Shipyard for Astro-Marine, Inc., cruises at 15 miles per hour as it enters the Gulf of Mexico at Galveston. Its tandem GM Detroit Diesel engines deliver 2000 horsepower.

Mangone Shipbuilding, purchased by Stewart & Stevenson in 1968, built work boats for the oil industry.

Mangone Shipbuilding Acquired

Mangone Shipbuilding Company, formed in June 1964 by Ivan Mangone, developed a split vessel design for building ships that spurred Stewart & Stevenson engine sales. The all steel, deep V, double chine hulls were built in halves, then welded together along the center line to form a complete, developed hull. One half measured 150 feet, the other 125 feet. The ships were powered by Stewart & Stevenson engine packages. Crew boats for offshore drilling rigs, shrimp boats, tugs and other work boats were offered in complete packaged deals by Stewart & Stevenson.

In 1968, Stewart & Stevenson acquired Mangone Shipbuilding Company to build "Packaged Boats" for use in the oilfields of the Gulf of Mexico. Among those was a 190-foot, 3200 horsepower offshore supply boat delivered to Briley Marine Service to operate initially in the Gulf of Mexico. Prime power was furnished by two 16V-149TI GM Detroit Diesel engines. Ships auxiliary power was furnished by two 125kW units powered by two 8V-71 GM Detroit Diesel engines. The boat also was equipped with a 300 horsepower Murray and Tregurtha bowthruster powered by a GM Detroit Diesel 8V-71 engine. The new ship was equipped with the latest navigational electronic equipment and had quarters for 22 men. It had racks for hauling drill pipe and compartments for wet and dry cement and wet and dry mud. The boat was capable of speeds up to 13.5 knots.

Mangone Ships Ply the North Sea

By the 1970s ships constructed under Stewart & Stevenson's "Packaged Boat Plan" were plying seas around the world, and Mangone Shipbuilding established a precedent in international shipbuilding in 1971 with the construction of the first offshore oil supply boats for a Norwegian firm that was not built by a

Norwegian shipyard. The discovery of oil beneath Norway's continental shelf necessitated the beginning of the offshore supply boat use in Norway.

The first order from a Norwegian firm was for two vessels for N.R. Bugge Supply Ships (BSS) A/S of Tornsberg, Norway. The *Sea Pearl* and its sister ship, the M/V *Sea Explorer* initiated the offshore, oil supply boat industry in Norway.

Stewart & Stevenson then received an $8 million order from Wilhelmsen of Oslo, Norway for four offshore service vessels.

Mangone Shipbuilding built four super supply ships for Fred Olsen of Oslo: two 185-foot vessels, the *Sea Bruse* and *Sea Bure*, and the 213-foot tug supply vessel *Sea Brute* and its sister ship, the *Sea Brisk*. All four vessels worked in the North Sea from Norwegian home ports.

For Bernhard Hanssen & Co., the fourth Norwegian ship owner to order vessels from Stewart & Stevenson, Mangone Shipbuilding built three tugs: *Musketeer 119, Musketeer 120* and *Musketeer 121*. Delivered in 1976, each had a cruising range of approximately 7000 miles. Two Detroit Diesel marine engines gave the vessels a cruising speed of 15 knots and 100 tons bollard pull.

The *Sea Pearl* was the first vessel built by Stewart & Stevenson to service Norway's offshore oil industry.

The *Musketeer Fighter,* one of three tugs built by Mangone Shipbuilding for Bernhard Hanssen & Co. of Norway, had a cruising range of approximately 7000 miles.

Diesel Generator Sets for Ships

Ingalls Shipbuilding Corporation in Pascagoula, Mississippi, contracted with Stewart & Stevenson to furnish 350kW Diesel generator sets to be used in five new cargo-passenger ships under construction for the American President Lines and the U.S. Maritime Administration. The five advanced automated cargo liners, a $68 million ship replacement program, ranked among the largest and fastest general cargo-carrying ships then operating under any flag. They were scheduled for use in American President Lines' Atlantic Straits service to carry cargo and passengers between the east and west coasts of the U.S. and to ports in the Orient and Southeast Asia — the longest American flag service in terms of nautical miles.

Stewart & Stevenson completely engineered and assembled the five 350kW Diesel generator sets, powered by GM Detroit Diesel 16V-71N engines, for installation aboard the vessels. The units, which were skid mounted and incorporated brushless alternators with static regulators, were both Coast Guard and ABS approved.

In 1966 Stewart & Stevenson Models 4GD-40 single phase 40kW sets and Model 6GD-100 three phase 100kW sets were furnished to Geotech Division of Teledyne Industries for use on ocean-going seismic exploration vessels. The units, mounted permanently on an outside forward deck of the firm's seismic exploration ships, were constantly exposed to wind, rain, sun and sea water salt spray. They furnished electrical power for Geotech's Marine Seismic Profiling Equipment (SSP) used around the world by Geotech in charting and gathering marine and geological data about the composition of the earth underneath the seas.

A marine generator set.

Project Mohole

The largest integrated well logging unit ever built left Stewart & Stevenson's Houston assembly plant in February 1966 for Schlumberger Well Surveying Corporation's testing lab. The unit would play a key role in the world's most ambitious drilling effort, Project Mohole. Stewart & Stevenson built and equipped the logging unit with everything except special electronic well-logging equipment that had to be added later.

General Motors Detroit Diesels furnished power for all three pieces of equipment. A Model 16V-71N of 564 horsepower furnished power for the main logging units and a Model 4-71N powered the auxiliary winch.

The two logging units and the service winch were part of the operational facilities to be used in drilling the world's deepest hole at a point in the Pacific Ocean about 115 miles north of Maui. The hole was to be drilled to a depth of 31,000 to 35,000 feet below the ocean surface from a floating platform in water approximately 14,000 feet deep. The primary objective of Project Mohole was to sample all layers of the earth's crust and the underlying mantle.

The completely assembled logging unit stood 18 feet high, 23 feet long and 9 feet wide. Completely outfitted, with the full length of cable wound on the winch drum, the unit weighed about 65,000 pounds.

The integrated well logging unit Stewart & Stevenson built for Project Mohole in 1966.

Military Contracts Fuel Earnings

For the Army

In 1963, Stewart & Stevenson was awarded a $3.7 million contract for 636 precise power Diesel driven generator sets for the U.S. Army. The contract called for delivery to begin in 1964. The new units, produced in the Houston plant, were to be exact duplicates of the Stewart & Stevenson Model 52-300, 45kW, 400 cycle precise power generator sets. The units supplied precise power for electrical systems used in space, communications and missile programs.

In 1964, 300 of the 45kW sets were delivered to the U.S. Army to provide ground support for the Pershing missile at U.S. and NATO military sites around the globe. Approximately 400 Model 52300 Stewart & Stevenson units were built.

For the Air Force

The following year, the U.S. Air Force awarded Stewart & Stevenson a $5.5 million contract for heavy duty engine generator sets. The contract called for the design of three different prototypes, with a fourth one to be added later. The prototype had to be built, tested and production models underway within 90 days.

Stewart & Stevenson designed, engineered, built a prototype and, on schedule, began delivery within 90 days of the issuance of the contract. The sets were rated at 30kW, 60kW, and 100kW for use under conditions ranging from minus 60 degrees to plus 135 degrees F. Known as the EMU family of generator sets, they were destined for general purpose base operations all over the globe. The units were powered by GM Detroit Diesel Model 3-71, 6-71, and 12-71 engines depending on the output requirement.

The EMU 16/E 60kW generator set built for the U.S. Air Force.

This Bomb Loader, built for the U.S. Air Force, picked up bombs and held them in the proper position for loading the bomber, which launched them into space,

For the Army

As Stewart and Stevenson completed the Air Force contract, a new $6.5 million U. S. Army contract with St. Louis Mobility Equipment Center called for 930 additional generator units of Model 52300 45kW 400 cycle precise power Diesel generator sets. They were to be delivered from late 1966 to mid-1967.

For the Navy

In 1966 delivery began on emergency generator sets especially designed for use on new Navy auxiliary ships. The company's contract from The Quincy Division of General Dynamics Electric Boat Division was to supply 500kW emergency Diesel-powered generator sets to be used for auxiliary power in four ammunition supply ships and two replenishment oil supply ships. All six generator sets were powered with tandem 12V71N GM Detroit Diesel engines with heat exchangers cooling. These engines were directly linked to a 500kW AC generator.

Under another contract, Stewart & Stevenson in 1968 delivered ten specialized high voltage engine generator sets to the U.S. Navy Bureau of Yards and Docks for advanced base application. These 500kW units were extremely light weight for their size, very compact in relation to the power output and weatherproofed for outdoor operation in any climate. Produced under contract with the U.S. Mobility Command, they were used

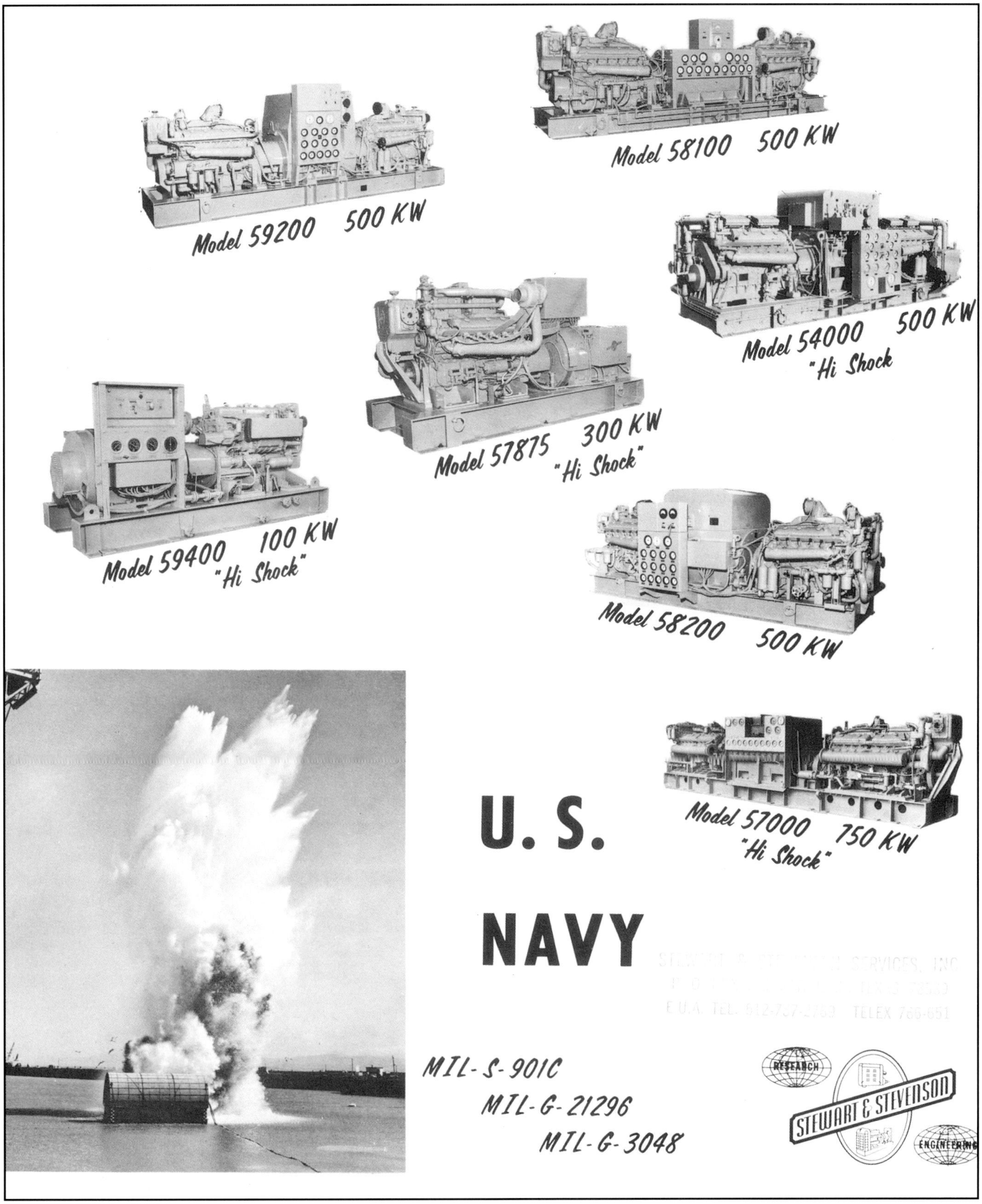

A new 300kW military generator set undergoes "high shock" testing (lower left). The combat generator sets, the first of their kind, were designed to withstand the severe shock of combat at sea.

by naval construction battalions when they moved into under-developed areas to set up operations. The generator sets served as prime power plants and operated with several units in parallel.

In 1968 Stewart & Stevenson expanded its capabilities in the field of "high shock" electrical power generator equipment for the U.S. Navy with a new, 300kW unit thoroughly proven in barge-explosion tests. The contract called for production of 14 new 300kW "high shock" units, all destined for naval landing craft. The combat designed generator sets, the first of their kind, were designed to withstand the severe shock of combat at sea. They were the first accepted as meeting military MIL. S-901C specifications.

Stewart & Stevenson engineers used a turbocharged General Motors Detroit Diesel Model 12V-71 engine driving a 300kW generator designed by Stewart & Stevenson for this specialized service. The unit was capable of operating at 10 percent overload at 1800rpm, 450 volt, 60 cycles. It was also equipped with a newly designed high shock voltage regulator and static exciter built by Stewart & Stevenson. Each unit was designed for auto start in the event of loss of regular shipboard power. Special heavy duty batteries provided power for electric starting and the operating frequency was quickly reached and maintained.

Constructed to meet military "maximum quality requirements" for supplier identification, the prototypes were tested aboard a floating barge by the Navy to qualify it. The test consisted of detonating explosives of varying severity at different distances from the barge while the unit was operating. When the unit continued to operate without damage under severe tests, it was certified as an acceptable "high shock" unit.

The U.S. Navy purchased these for emergency backup power aboard a variety of craft. Landing craft were usually exposed to the most severe type of combat operations and relied heavily on dependable equipment. The units were shipped to Ingalls Shipbuilding Company, Lockheed Shipbuilding and Construction Company, and General Dynamics Quincy Division for installation into Navy ships. They were to be delivered from late 1966 to mid-1967.

For the Air Force

In 1968, the Air Force placed a third order for Mobile Power Houses, special Diesel generators mounted in trailers. Each unit was a completely self-contained electric power plant equipped with two 150kW, 60 cycle AC Diesel generators. Power for each generator was supplied by a Model 8V-71 GM Detroit Diesel engine. A special Stewart & Stevenson designed and built panel for switch gear and operating instruments was mounted in the van between the two generator sets. The vans were used to provide dependable electrical power sources for the Air Force in various field communications warning systems and other applications where electric power was required.

Mobile Power Houses for the U.S. Air Force provided electrical power in the field.

Consolidating Power Industry Position 1970 - 1980

A Company Without Limits

"The world of Stewart & Stevenson has no limits," noted an article in the June 21, 1970 *Wall Street Journal*. "From the power plant at the Johnson Space Center, which backs up every space flight, to the palace of a prince in Saudi Arabia, and even a house of ill-repute in Nuevo Laredo, Mexico, Stewart & Stevenson generator sets are on the job."

Close to home, Stewart & Stevenson was helping the city of Lubbock, Texas get back on its feet following the May 11 West Texas tornado. Stewart & Stevenson generator sets, sent from the company's Amarillo, Odessa and Houston plants, were providing essential power no longer available from the city's badly damaged electric plants. Three months later, the company performed the same community service in Corpus Christi following hurricane Celia, which came ashore August 3.

Generator sets for prime, uninterruptable and standby power remained a key component of the company's operations, as did innovative Diesel power for oilfield applications, irrigation and pumping, airline ground support and "Packaged Boats." Along with worldwide parts and service, they kept Stewart & Stevenson positioned as the world's largest distributor of Diesel engines in 1970.

Charles L. Ward, then vice president and general manager of Stewart & Stevenson Services, said the 1970 sales volume, which exceeded $40 million, was indicative of an "aggressive, determined organization during a year that left a lot to be desired in the nation's overall economic climate." Diversification initiated two decades earlier was paying off. Before this decade ended, the company's annual sales would exceed $100 million.

A Stewart & Stevenson generator provides backup power for the Houston Police Department.

"Sparky," the spark-ignited gas generator, applied in refineries, petrochemical plants, pipelines, construction, oilfield offshore platforms, general industrial, standby power and other general purpose uses.

Product Innovations

Turbine engines introduced included the Allison Series 501K gas turbines for industrial-marine-petroleum applications in single gas turbine packages ranging from 2000 to 5000 horsepower and multiple gas turbine configurations up to 20,000 horsepower. A new gas compressor package powered by the 7000 horsepower Detroit Diesel Allison 570-K Industrial Gas Turbine also was introduced.

Generator advances included a new line of economically priced natural gas generator sets in sizes from 55kW to 150kW. The spark-ignited gas engines were applied in irrigation, refineries, petrochemical plants, pipelines, construction, oilfield offshore platforms, general industrial, standby power and other general purpose uses. Also, the company's main line of Detroit Diesel and Electro-Motive Division-powered AC Diesel and dual fuel engine-driven generator sets was expanded.

Other introductions included a Gater aircraft gate tractor designed to move larger aircraft, and firefighter engines in six vertical and six horizontal models.

Diversification, Expansion Continues

In 1970 the Thermo King Division was formed to sell and service Thermo King refrigeration equipment to the transportation industry.

In 1972 Stewart & Stevenson formed the International Switchboard Corporation (ISC), primarily

Three Allison 501K gas turbines provide power for this Penrod Drilling Company offshore drilling rig.

to supply the ever growing requirements of the powerhouse type switchgear for the Engine Division.

In 1973 Stewart & Stevenson became the authorized distributor of Allison transmissions in the Southwest.

From Olin Corporation, Stewart & Stevenson acquired manufacturing and patent rights for Pow-R-Quik Air Starters for industrial equipment, which were handled by Stewart & Stevenson's Specialty Products Division headquartered in Houston.

The London office of Stewart & Stevenson, managed by Logan Knapp, was opened in 1973 to take care of the company's many European customers in the petroleum industry.

In 1974 the Garrett Corporation appointed Stewart & Stevenson Services, Inc. as a major outlet and packager of Garrett industrial gas turbine engines.

In 1975 a joint venture, Stewart & Stevenson de Venezuela, was formed to operate as sole distributor for GM Detroit Diesel engines and Allison transmissions in Venezuela, South America.

In 1978 Stewart & Stevenson purchased the Dobbs Detroit Diesel Distributorship in Denver, Colorado, which became S&S Power, a wholly owned subsidiary, adding Colorado, parts of Wyoming and Nebraska to the company's territory.

Saudi Diesel, a joint venture with a Saudi partner, was formed in 1978 as a Detroit Diesel-Allison distributor for the Kingdom of Saudi Arabia.

Also in 1978, Stewart & Stevenson entered into a contract with Riley Beard, Inc., Shreveport, Louisiana, for the exclusive worldwide sales and service of Maxim Packaged Desalinators to the petroleum industry.

In 1979 Stewart & Stevenson formed International Electric Corporation (IEC), licensed to manufacture and sell Delco designed generators.

The Company Reorganizes and Goes Public

The allied companies of Stewart & Stevenson — Stewart & Stevenson Services, Inc.; C. Jim Stewart & Stevenson, Inc.; Alemite Company of the Gulf Coast; C. Jim Stewart & Stevenson, Inc., Truck Lease Division; C. Jim Stewart & Stevenson Southwest Company, Materials Handling Division; Stewart & Stevenson Distributing Company, Oilfield Specialty Division; and Stewart & Stevenson Distributing Company, Paint and Automotive Division — were reorganized in 1972 with Stewart & Stevenson Services, Inc. becoming the parent company.

A corporate structure chart in the 1974 annual report revealed two branches formed under Stewart & Stevenson Services, Inc., originally the Engine Division: C. Jim Stewart & Stevenson, Inc., which included the Material Handling Division, Truck Lease Division and Alemite Division, and the Engine Division/Stewart & Stevenson International Corp., under which fell the International Switchboard Corp., Machinery Acceptance Corp., Stewart & Stevenson Realty Corp., Stewart & Stevenson S.A. de C.V., Stewart & Stevenson Oil Tools, Inc., and Mangone Shipbuilding Company.

Seven Divisions Formed

For practical purposes, the company was arranged in seven divisions: Engine Division, Oiltools Division, Mangone Shipbuilding, Truck Lease Division, Alemite Division, Material Handling Division, and International Switchboard (ISC).

As it had since 1903, Stewart & Stevenson

Ross Stewart, right, presents his son, C. Jim Stewart II, a 25-Year Service Award (watch).

C. Jim Stewart II, President (1973-1976), CEO (1985-1991), Chairman of the Board (1976-2001), Chairman of the Board Emeritus (2002).

remained tuned to the basic industries it started: power and transportation. Chairman Emeritus Ross Stewart, who strongly believed that Stewart & Stevenson's base and future should be directly tied to the application of power, was instrumental in the formation of each of the divisions, which had power as a common thread. The company's guiding force, Ross Stewart continued to impact Stewart & Stevenson's direction until his death on May 28, 1975, at the age of 78.

C. Jim Stewart II, President

In 1973 J. Donald Stevenson became Chairman of the Board, and C. Jim Stewart II, a third generation Stewart and grandson of the company founder, became President of Stewart & Stevenson Services, Inc. Under C. Jim Stewart II's direction, the company went public in 1976, offering 850,000 shares of Stewart & Stevenson Services, Inc. common stock at $17 per share. The company advertised its first public offering in the June 1976 *Wall Street Journal*.

"There was initially a lot of reluctance to take the company public," C. Jim II recalled during a December 1999 interview, adding that he based the decision to do so on three reasons: to obtain the necessary capital to grow the company; to get value for the family stockholders who needed income from their stock; and to bring objectivity to the board and broaden the focus of the company vision.

"The Stewart and Stevenson families had grown," Stewart explained. "Individual family members had a lot of net worth in the company, but no income from their portion of ownership. The company didn't pay dividends because it needed every dollar of profit to support its rapid growth. Family wise, it was a timely decision.

"We also wanted to enable our employees to have an opportunity to build some net worth outside of their income so they would have something to rely on when they retired. After going public, we offered employees the opportunity to buy stock and matched their purchases up to a certain amount. That way, they could improve their net worth.

"But the main reason was to have more capital to expand the business. The banks were very good to loan us what we needed, but they had more money in the company than the family did, and that was a little unhealthy. They told us we needed more capital. We had no option but to go public," Stewart concluded.

Today, Stewart & Stevenson common stock is traded on the NASDAQ Stock Market under the symbol: SSSS.

Corporate management of Stewart & Stevenson Services, Inc. during C. Jim Stewart II's presidency included Joe W. Manning, executive vice president; Lewis Edwin Ball II, vice president and treasurer; Paul Clifton Koomey, vice president and general manager, Oiltools Division; Robert M. Hardy, vice president and general manager of C. Jim Stewart & Stevenson, Inc. Alemite Division; J. Carsey Manning, vice president of sales, Engine Division; Thomas H. Langham (another grandson of the company founder), vice president of manufacturing, Engine Division; Leamon Tankersley, general manager, C. Jim Stewart & Stevenson, Inc. Material Handling Division; Richard Noble Conolly, vice president branches, Engine Division; Charles L. Ward, vice president and general manager, Engine Division; Ronnie Latham Pruett, general manager C. Jim Stewart & Stevenson, Inc. Truck Lease Division; Don Louis Godeau, vice president and general manager, Mangone Shipbuilding Company; and John Herbert Shaw, general manager, International Switchboard Corporation.

Engine Division

The Engine Division, the company's largest, continued to engineer and fabricate anything that utilized a Diesel or turbine engine, and provided 24-hour, 365-days-per-year parts and service back-up. A 79,000-square-foot central parts distribution warehouse, housing an inventory of $20 million in service parts, utilized automated handling equipment and an on-line computer system to supply dealers and branches worldwide.

Principal markets for Stewart & Stevenson engines were the petroleum industry, including drilling, pipeline and construction, production and exploration; the marine market, including commercial fishing, workboats and pleasure craft; irrigation; commercial airlines, including electrical ground power units, air start units for the large jets, and repowering; power plants, including standby and emergency power generator sets for nuclear power plants, larger prime power units for elec-

The 70,000 square-foot Central Parts Distribution Center in Houston housed more than $20 million in parts to supply the company's dealers and branches worldwide.

First Fourth Generation Stewart Joins the Company

Garth C. Bates, Jr., grandson of Ross Stewart and nephew of C. Jim Stewart II, became the first of the fourth generation Stewarts to work full-time for the company. After obtaining a degree in Finance from the University of Texas at Austin, Garth joined the Engine Division in the summer of 1971, beginning at the Corpus Christi branch under Dick Conolly and Roy Markwardt. In 1973 he moved to Houston to serve under Carsey Manning and Bill McFarland as a contracts administrator for the DD963 class ship service generator and distribution switchgear program — a $9 million switchgear contract and $40 million generator contract. Garth served as a contracts administrator until 1977, when he became manager of the San Antonio branch. In 1981 Garth moved to Denver, Colorado as President of S&S Power, which had branches in Denver and Rawlings, Wyoming. In 1991 he returned to Houston to become vice president in charge of the distribution business. Garth resigned from Stewart & Stevenson to pursue other opportunities in April 2000.

tric power generators, and hospital standby; and U. S. Government sales, including the ship service, turbine-driven generator sets for the Spruance Class, DD963 destroyer program.

Principal supplier for the Engine Division was the Detroit Diesel Allison Division of General Motors Corporation.

Oiltools Division

The Oiltools Division, Stewart & Stevenson's second largest operation, benefited directly from the rapidly increasing world-wide oil exploration program. Primary product was the Koomey blowout preventer closing system, which had become a generic term in the industry for all BOP control systems.

The Koomey Closing Unit.

Both economic and environmental concerns dictated petroleum exploration practices. The products of the Oiltools Division were designed to prevent oilwells — both onshore and offshore — from creating expensive, environmental hazards.

Stewart & Stevenson developed new and improved controls for the innovative new drill ships that were capable of drilling and completing an oil well in more than 6,000 feet of water. The control line operating the blowout preventer was reduced from a cable the size of a man's arm to that of a silver dollar. Also, new technology was developed utilizing electronic versus hydraulic controls.

Oiltools manufacturing facilities were consolidated in a new 220,000-square-foot plant that opened in 1975. The division was sold to N.L. Industries in 1978. In 1986, the company re-entered this market as S&S Oilfield Products Division.

C. Jim Stewart III Joins Company

C. Jim (Jimmy) Stewart III, son of C. Jim Stewart II and fourth generation member of his family, joined the company as a sales trainee in the Oiltools Division in January 1972 after graduating from Texas Christian University with a B.B.A. in Marketing. He worked in the Service Department for a year and traveled all over the world working on oilfield equipment. In 1973 he moved to Subsea Sales, and later became manager of Subsea Sales. When the Oiltools Division was sold, Jimmy moved to the Oilfield Products Group. In 1981, he became manager of Petroleum sales at the Houston Branch, then became sales manager of that branch in 1985. In 1987 he moved to corporate as the new Vice President of Marketing. Today he heads the Marketing Services Group which supports the company's overall sales and marketing efforts worldwide.

Mangone Shipbuilding specialized in the construction of tugs and supply ships to meet conditions of all oceans and waterways.

Mangone Shipbuilding Division

Mangone Shipbuilding Company specialized in the construction of tugs and supply ships to meet conditions of all oceans and waterways. In spite of spiraling inflation and a steel shortage, Mangone Shipbuilding delivered five highly sophisticated 185-foot workboats to a Norwegian shipping company in 1974, and keels were laid for a new 213-foot tug/supply vessel. These powerful boats, powered by twin screw, 3500 horsepower EMD Diesels driving controllable pitch propellers for greater flexibility under variable conditions, were for use under the savage conditions of the North Sea.

The company also built a new type 150-foot ocean tug for operation in the North Sea. These vessels, built as Class C ice breakers, were used to tow vessels as well as service offshore drilling. Mangone Shipbuilding was the only American shipyard utilized by Norwegian ship owners to build their vessels.

As Stewart & Stevenson consolidated its position in the power industry, Mangone Shipbuilding was sold to Swiftships in 1981. The company continued to market package boats with hulls built by other shipyards.

The EMD engine supplied propulsion power for supply ships.

Stewart & Stevenson's Truck Lease Division opened two new truck lease facilities in Houston in 1974, and had a fleet of 450 trucks. The division was sold near the end of the decade to ABC Leasing.

Thermo King became a separate division after the Truck Lease Division was sold.

Truck Lease Division

Although high fuel costs and fuel shortages in the early 1970s adversely affected the Truck Lease Division, it continued to serve customers requiring large multi-unit fleets of trucks as well as those requiring a single unit on an occasional basis. The company's truck fleet numbered more than 450, and the fleet of company cars put under its management brought the total number of vehicles managed to approximately 600. In 1974, two new truck lease facilities were opened in Houston to better serve customers.

When the Truck Lease Division, which also sold and serviced Thermo King truck refrigeration units, was sold to ABC Leasing in the late 1970s, Thermo King became a separate operating division.

Richard Ross Stewart Joins Company

Richard Ross (Rick) Stewart, son of C. Jim Stewart II and fourth generation Stewart, received a B.B.A. in Finance from the University of Texas in May 1972. After joining the Truck Lease Division in November 1972, he developed a computer program to track costs and profitability of each truck. Then, because the division had a mostly local clientele, he convinced the company to give up the Hertz, Ryder and Rollins franchises and become simply S&S Truck Leasing. Both decisions aided the division's growth and increased profits. When Dave Klager, who ran the airline ground support business, suffered a heart attack, Tom Langham and Joe Manning convinced Rick to become the sales rep for the airline ground support business. After three years in airline sales, he was put in charge of airline ground support equipment manufacturing. In 1980 he became manager of all manufacturing at the Harrisburg plant. In 1986, Rick was made Vice President of the Gas Turbine Division, charged with acquiring, redesigning, and staffing the Jacintoport facility. Under his direction, the Gas Turbine Division grew from a $20 million business in 1986 to a $650 million business in 1995 when it was sold to General Electric. Today, Rick is President and CEO of General Electric Aero Energy Products, a GE Power Systems Business.

The Alemite Division expanded service to a wide variety of companies with worldwide operations before it was sold back to Stewart-Warner Corp. in mid-1970.

Alemite Division

A distributor of lubricants and allied equipment, the Alemite Division served a wide variety of companies which had expanded their operations world-wide. As a result, the regional distributor reaped the benefits from such things as increased drilling in Alaska to large power plants under construction in the Mideast.

Imaginative technical approaches continued to expand the division's base. For example, to meet EPA requirements for a new nuclear power plant, radical innovations were required for fueling and lubricating construction equipment. A system was developed that would fuel large equipment at the rate of 150 gallons per minute. A similar system for lubricating large equipment allowed a complete change of oil in a matter of minutes rather than hours and was accomplished while the equipment remained at its work location.

By 1979 the Material Handling Division ranked among the top five Hyster distributorships in the nation.

The Alemite franchise was sold back to the Alemite Division of Stewart-Warner Corporation in the mid 1970s.

Material Handling Division

The Material Handling Division, which began in 1959, handled both new and remanufactured models of Hyster equipment. A new product introduced by Hyster in the early 1970s for the Texas agriculture industry was a large crane, referred to as a "Cherry Picker." Capable of handling 36,000 pounds, it found a ready market in the citrus and truck farms of the Texas Rio Grande Valley. A new service center was opened in Harlingen, Texas, to serve this new industry.

In 1979 the Material Handling Division of Stewart & Stevenson ranked among the top five Hyster distributors in the United States in terms of total dealer sales. The full line of Hyster forklifts included lifting capacity ranges from 2,000 pounds to 80,000 pounds. Forklifts were available with gasoline, electric, liquefied gas and Diesel power.

The Material Handling Division's 1979 sales had increased more than 50 percent, accounting for nine percent of corporate sales. This rapid growth was attributed to the division's young, aggressive management team and an economically strong South Texas distributor territory. The Material Handling Division's new 790,000 square-foot Houston headquarters opened early in 1980. Branches were located in San Antonio, Galveston, Beaumont, Corpus Christi, and Harlingen, Texas. Today the Hyster industrial equipment group covers the entire state of Texas.

International Switchboard (ISC)

Since its beginning in the Diesel business, Stewart & Stevenson had fabricated various types of electric switchgear, switchboards for its generator sets. In 1972 the company formed the International Switchboard Corporation (ISC), which became a major supplier of electrical control and power switchgear for power generation and distribution. As Stewart & Stevenson developed Diesel and turbine engines for newer applications, ISC developed switchgear and control systems for those applications.

SCR control systems, developed primarily to provide Diesel electric power for drilling rig applications, were used in various applications where variable speed DC motors were required; for example, in ships, locomotives and mining machinery. ISC also was active in the design and manufacture of electrical systems for nuclear power plants, public utilities and military applications. Stewart & Stevenson, through ISC, offered the advantage of complete and single responsibility, from the Diesel prime mover through the alternating current generator, the SCR system, AC switchgear, and interface.

ISC served the switchgear requirements both of the Engine Division and unaffiliated customers throughout the world. The modern facility constructed in 1972 exclusively for the manufacture of electrical switchgear was continually being expanded to meet the need.

ISC was sold to Equus Power Systems in June 1983.

Expanding Branches

Stewart & Stevenson now had seven branches and 20 major dealers in Texas and New Mexico to handle sales, parts and service for its Diesel and turbine power applications. Major expansions were underway at many of them.

Left: Don Wallin, E.E. "Slim" Childress, Jim Murray and Paul Gosnell (*left to right*) break ground for the new Houston branch off I-10 in Houston.

C. Jim Stewart III Recalls Ross and C. Jim Stewart II's Views on Hiring Family To Work For Company

When he asked for a job at Stewart & Stevenson, C. Jim Stewart III recalled that his father told him, "I can't guarantee you anything other than a job opportunity when you finish college."

"My grandfather," C. Jim III recalled about Ross Stewart, "gave everyone a book on the pitfalls and dangers of nepotism. He drove home the idea of how nepotism can be your strongest asset, or it can be your greatest threat, if done at a cost of others.

"As family members employed in the company, we were all taught by our Dad that respect must be 'earned' and would be given! We had to earn our positions, and prove ourselves on our own. We had to work harder because people expected more from us," C. Jim III recalled.

It was also C. Jim II's philosophy that his sons, if they did come to work for the company, should each work in different areas of the company.

"Keith and Don Stevenson were both working in the Engine Division, and Garth Bates also began his career in Engine operations," C. Jim III said. He was sent to the Oilfield Division, Rick to the Truck Lease Division and David to the Corporate Office. Each began as a trainee, learning different aspects of the business, and working his way up.

"Stewart & Stevenson," he continued, "is a company with deep roots and dedicated to loyal employees. It is a fraternity of people with a common goal under a common roof called Stewart & Stevenson. We must never, ever forget that it is not the Stewarts nor the Stevensons that are responsible for the company being where it is today, but the thousands of employees and their families that have made the company what it has become over the years. The guy in the shop wearing a uniform shirt with the Stewart & Stevenson patch is just as important as the guy in the white shirt in the corporate office. The dedication and commitment of the people who work for Stewart & Stevenson has always been the company's greatest asset.

"Would we be more successful if we had changed our philosophy from a family-oriented company?" he pondered, then answered his own question, "Maybe, but at what cost to our 'core values'?"

David R. Stewart

David Rial Stewart, youngest son of C. Jim II and fourth generation Stewart, graduated from the University of Texas with a degree in Finance, and began working full time in the corporate office in July 1973. He started in the accounting department under Bill Pieper, working in payroll and general accounting. Just before the company went public in 1975, he took charge of corporate real estate, which was owned by family members under a separate company called Bayou Building Company. When the company went public, Bayou Building was merged with Stewart & Stevenson to form Stewart & Stevenson Realty Corp. In 1976 David became involved in investor relations and worked in financial planning and analysis. In 1999 he became treasurer of Stewart & Stevenson Services, Inc.

Tom Langham IV

Tommy Langham, grandson of Aileen Stewart Langham and son of Thomas H. Langham III, is also a fourth generation member of the Stewart family. He began working full time at Stewart & Stevenson in 1976 and continues today as a loyal, dedicated employee in the Fabrication area of the Houston branch of Stewart & Stevenson.

Energy Shortages Fuel Demand for Diesel Power

As energy shortages brought rapid escalation in the cost of energy, more and more companies turned to Stewart & Stevenson for Diesel power systems to meet their energy requirements, and Stewart & Stevenson worked to achieve greater fuel efficiency and economy of operation in virtually all its power applications.

Stewart & Stevenson experienced the rapidly rising demand for Diesel power systems in a number of ways: higher sales of emergency generator sets as the result of brownouts and blackouts; increasing Dieselization of "medium duty" truck fleets; rising demand for Diesel equipment by the petroleum industry as exploration and drilling activity stepped up; sharply increased sales to airlines of more economical Diesel-powered ground support equipment; and increasing use of Diesel-power systems as primary and secondary power sources in emerging nations.

To take advantage of these opportunities, Stewart & Stevenson embarked upon the largest expansion program in its history, adding facilities to increase production, broaden its geographic territory, increase product development capability, and enable the company to further penetrate major markets. Experienced, well-qualified people vital for growing the business also were put in place.

Joe Manning, President

In 1976, C. Jim Stewart II became Chairman of the Board and Joe Manning became the first non-family member President and CEO of Stewart & Stevenson Services, Inc. Hired by Ross Stewart as Diesel salesman in 1938 at age 19, Manning became general manager of the Engine Division in 1939, vice president and general manager of the Engine Division in 1946, and executive vice president of Stewart & Stevenson Services, Inc. in the early 1950s.

Under Manning's leadership, the company progressed from primarily an automotive supply and repair business to become the "World's Largest Distributor of Diesel Engines." He managed the company's sales and production of Diesel-powered products during World War II and, after the war, set up the Engine Division's branches and dealers and initiated the annual sales meetings. He was instrumental in the sale of the Truck Lease and Oiltools divisions, and the acquisition of the Dobbs Detroit Diesel distributorship, the Detroit Diesel distributorships for the Kingdom of Saudi Arabia and all of Venezuela, and the formation of Stewart & Stevenson de Venezuela. Joe Manning retired in 1985 after 47 years of loyal and dedicated service to the company. He continues to serve as a corporate advisor.

Joe Manning, President & CEO (1976-1985).

A Diamond Anniversary

In 1977, Stewart & Stevenson Services, Inc. celebrated 75 years of innovation. The company had grown from a two-man "pardnership" to an international public corporation employing more than 3,000 people.

Facilities Expansion

Expansion of the Engine Operations created a need for greater parts and service, so Stewart & Stevenson constructed a new $7 million, 144,000-square-foot Parts & Service Center on a 19-acre tract in Houston. Put into operation in early 1980, it enabled the Engine Operations to handle two to three times more Diesel and turbine repair business and freed up much-needed space at the Houston manufacturing and assembly plant for additional power systems production.

Stewart & Stevenson parts remanufacturing, begun by the company in 1956 in San Juan, Texas, expanded with the San Antonio remanufacturing center that went into full operation during 1978. That operation remanufactured Diesel parts, Detroit Diesel engines, automatic transmissions, and engine components.

With the 1980 opening of the new 70,000 square-foot Houston headquarters facility for the Material Handling Division, and the addition of two new branches in Austin and Wichita Falls, the $30 million capital expenditure program begun in 1977 was completed.

Stewart & Stevenson de Venezuela S.A., formed in 1975, was headquartered in Maracaibo with a branch office in Caracas.

New Vice Presidents Added

In 1978 Charles L. Ward, executive vice president and general manager of Stewart & Stevenson Services, announced the appointments of three new vice presidents: C. LaRoy Hammer, Miles McInnis and Ray Loden.

C. LaRoy Hammer, who had served 21 years in sales and management positions, became Director of Government Contracts.

Miles McInnis, who had been manager of the Dallas branch for 14 years, was named Vice President and Director of the Dallas Division of Stewart & Stevenson. The Dallas operation was made a division as part of the company's long-range expansion program.

Ray Loden, who began his career as a mechanic at Stewart & Stevenson in 1947 and had since served in positions of sales and management in the marine division, became Director of Marine Sales in Houston.

Stewart & Stevenson de Venezuela S.A.

Stewart & Stevenson de Venezuela S.A., headquartered in Maracaibo, was formed in 1975 to sell and service Detroit Diesel engines and products in South America. Under an agreement with a Venezuelan company, Stewart & Stevenson Services furnished the technical services, expertise and management necessary to perform the manufacturing, application, assembly and distribution of Detroit Diesel engines and engine driven products throughout Venezuela. Stewart & Stevenson owned 50 percent of Stewart & Stevenson de Venezuela. The affiliate also assembled generator sets and operated a complete remanufacturing center for reclaiming engine parts and components.

Headquarters was a 22,000 square-foot facility located in Maracaibo, where a major expansion was completed in 1979. A branch office was located in Caracas. Stewart & Stevenson developed a strong dealer network in Venezuela, patterned after U.S. operations. Support dealers with sales and service assistance and training, parts & service availability plus an aggressive sales approach resulted in rapid growth for this South American company in all markets: oilfield, mining, construction, marine, utilities, trucking and agriculture. Sales of Detroit Diesel Allison parts or products in Venezuela more than doubled in the three years after its opening under Stewart & Stevenson management.

Stewart & Stevenson's success in Venezuela led to a managing partnership with Saudi Arabia for sales and service of Detroit Diesel Allison engines.

Saudi Distributorship Granted

The Detroit Diesel Allison Division of General Motors granted the sole distributorship for the Kingdom of Saudi Arabia for the sales and service of Detroit Diesel Allison engines and equipment to a joint venture in which Stewart & Stevenson was the managing participant. Stewart & Stevenson's success in Venezuela had been a key factor in GM's decision to back the company's entry into Saudi Arabia.

Saudi Diesel was 60 percent owned by Saudi citizens and 40 percent by Stewart & Stevenson. A 56,000-square-foot parts & service center constructed at Al Khobar in 1979 also contained facilities for assembly of generator sets, engine power units and small switchgear. Stewart & Stevenson's Houston plant provided the more sophisticated power systems.

Stewart & Stevenson invested $1.2 million in the joint venture and received 50 percent of the profits.

Power plants and other facilities in Saudi Arabia requiring Diesel power were built at a rapid rate as the country proceeded with industrialization efforts.

Rocky Mountain Acquisition

In July 1978, Stewart & Stevenson acquired, at a cost of $7.0 million, Dobbs Detroit Diesel, Inc., the General Motors Detroit Diesel Allison distributor franchise encompassing Colorado, most of Wyoming and part of Nebraska. The wholly owned subsidiary of Stewart & Stevenson began operating as Rocky Mountain Power Services, Inc., but became Stewart & Stevenson Power, Inc. in 1980.

Changing what had been an "off-the-shelf" engines and parts capability, Stewart & Stevenson Power upgraded its Denver headquarters' service facilities and began selling custom-engineered power systems. Some Diesel-powered systems were manufactured and packaged on site; the more sophisticated power equipment was built in the Houston plant. By 1980, the subsidiary was providing the entire line of Stewart & Stevenson Diesel and gas turbine-powered products, including parts and service.

Stewart & Stevenson Power added new territory and increased its sales force in the mining, irrigation and oilfield markets. The subsidiary established two branches in Grand Junction, Colorado and Rawlins, Wyoming.

Rocky Mountain Power in Denver became Stewart & Stevenson Power, Inc. in 1980.

Generators Serve Many Industries, Nations

For Nuclear, Other Power Plants

When the 840,000kW nuclear generating station in Russelville, Arkansas — the first nuclear power station in the Southwest — began operation in 1971, it had precise emergency backup power provided by Stewart & Stevenson. The company designed, engineered and assembled two 2,750kW emergency Diesel generator sets for the new Arkansas Power & Light complex.

"Provisions had to be made for the units to accept the full power load within 10 seconds after commercial failure of initiation of start sequence," said Logan Knapp, then manager of Stewart & Stevenson generator sales. "In addition, special alternators had to be supplied that would hold very close tolerances on voltage produced by the generating units while starting large motors."

The generating units also had to be capable of withstanding earthquake shock criteria as established by the Atomic Energy Commission.

Dallas Power & Light Company selected a 350kW standby generator unit for the 375,000kW Unit No. 1 at its Lake Ray Hubbard generating station. Designed to start automatically in case of normal voltage outage, Stewart & Stevenson emergency units furnished essential power for the air compressor used for plant instrument operation, for the turbine generator turning-gear motor and gear-oil pump, for emergency plant lighting, boiler feed controls, service fire water pump, plant elevator and numerous other motors required to operate the facility.

Stewart & Stevenson manufactured the first nuclear power plant installation of the split crankshaft Diesel engine-driven generator set. Three 3500kW emergency tandem Diesel generator sets were assembled for the Florida Power & Light Company for installation in the New Hutchinson Island Nuclear Station near Fort Pierce, Florida. Each unit consisted essentially of two GM Electro-Motive engines operating in tandem, powering 900rpm 3500kW alternators, and employed Woodward electrohydraulic speed and load-sensitive governors. Stewart & Stevenson also furnished the electrical control panels, fuel supply system, skid-mounted radiator cooling system, starting air system, and other required auxiliary support systems.

Stewart & Stevenson also supplied three nominal 3500kW tandem Diesel generator sets to Cincinnati Gas & Electric Company's Zimmer Plant in 1974.

The company produced two 2200kW emergency Diesel generator sets for Allied Gulf Nuclear Services (AGNS) for installation at an AGNS nuclear fuel processing plant in Bayville, South Carolina.

For Airlines, Airport Facilties

An uninterrupted power system developed from a computerized dynamic design approach by Stewart & Stevenson was installed at National Air Lines' new $17 million computer nerve center. The highly complex airline reservations system, located in Miami, was the most advanced in the industry and represented five years of extensive research and thousands of man hours. Because the sensitive computers require absolute and very precise uninterrupted power, National engineers selected Stewart & Stevenson to provide an electric standby power system.

Eastern Airlines' third generation Computer Center had a 2300kW Emergency standby Diesel generator set powered by an EMD Series 645 Diesel engine.

In 1972, Stewart & Stevenson provided emergency power equipment for Dallas-Fort Worth Regional Airport, the world's largest airport then being constructed under a master plan that extended to the year 2001. The company designed and produced standby emergency generators to support three of the most critical installations at the airport: the Airport Communications Center, the Central Utilities Complex, and the airfield runway lighting. A 450kW emergency generator powered by a Detroit Diesel 16V-71T Diesel engine was selected for emergency support of the communications facilities. A 225kW emergency generator powered by a GM Detroit Diesel engine Model 8V-71TI was selected for the utilities complex, and two 800kW emergency Diesel generators powered by the 16V-149T provided the required emergency lighting for the airfield runways.

Power for Remote Areas

Stewart & Stevenson delivered a new 2000kW GM Detroit Diesel-powered generator set for Empresa de Energia de Bogota, Colombia. Mounted in a van-type trailer within dimensions of highway limits, it was specially designed to be trailed over highways and roads with maximum portability. Its initial destination was the Chivor hydroelectric project located in a remote and rugged area 125 miles north of Bogota where it supplied power during the construction of the hydroelectric facility. Stewart & Stevenson engineers teamed with the company's production department to design, construct and deliver the unit dockside within 60 days after receipt of the purchase order, quite an accomplishment.

A $1 million-plus power plant was delivered in record time to Bechtel Corporation, agent for Bougainville Copper, a company formed to exploit a

Stewart & Stevenson furnished more than 16,000kW in mobile Diesel generator sets designed for transport throughout Colombia to support the Colombian Electrificacion program.

rich copper deposit in a remote jungle location on Bougainville Island. The 11,000kW EMD Diesel-powered generating plant provided power for the large construction camp built near the mine site to house construction workers, and also for two of the world's largest stripping shovels, which removed as much as 100 feet of overburden from the copper deposit in preparation for mining operations.

Under a $2 million contract from Instituto Colombiano Electrificacion (ICEL) of Bogota, Colombia, Stewart & Stevenson furnished over 16,000kW in mobile Diesel generator sets designed for transport throughout the country. Delivered in 1971, the units were placed at strategic locations throughout Colombia to support the Colombian Electrificacion program.

Generator Sets for COMSAT

Communications Satellite Corporation of America (COMSAT) obtained authority from the government of Saudi Arabia to install satellite receiver/transmitter stations at Riyadh and Jeddah. COMSAT contacted Stewart & Stevenson for four automatic Diesel generator units and switch gear for use as prime power at the stations. Two months after receiving the order, the complete equipment package was put on a plane to Washington, D.C. where it was transferred to a charter flight for delivery to Riyadh. Approximately 45 days later, the complete COMSAT earth station was in operation at 100 percent efficiency, and a second station became operational 30 days later. The automated and redundant power generation equipment operated continuously in the hostile environment of Central Saudi Arabia, in temperatures that often reach 120 degrees F. at midday and in high winds that keep dust and sand contaminants airborne at virtually all times.

Hi-Shock Generators for Navy Ships

A major contract to furnish specially designed Hi-Shock Diesel generator sets on nine revolutionary United States Navy general purpose assault ships was awarded to Stewart & Stevenson Services by Litton Systems, Inc. The generators were utilized in combination for the ship's service and standby power. Deliveries began in 1971 and extended through 1974.

Diesel Power in the Oil Fields

Stewart & Stevenson served nearly every segment of the petroleum industry: drilling, pipeline construction, production, exploration, and oil well servicing. The company's Diesel drilling engines, SCR systems, mechanical power transmission drives, turbine powered gas compression systems, generators and pumps were utilized by land rigs, offshore rigs and workover rigs. Its Diesel electric drilling system included Diesel-powered AC generators, DC motors, SCR controls and AC interface.

Workover Rig Development

Stewart & Stevenson overcame barriers of space and weight limits to provide power packages for five unique offshore platform workover rigs owned by the Houston-based Gulf Offshore Company. These rigs, based on the "Go-Getter" concept developed by Jim Armstrong of Gulf Offshore in 1970, consumed 60 percent less platform space and could be moved between platforms without derrick barges or crane boats.

The generator utility house, which furnished all electric power requirements for the rig, consisted of two Stewart & Stevenson 50kW, model 3-71N General Motors Detroit Diesel-powered, explosion proof generators, an air compressor with air receiver, and an explosion-proof panel board. The roof of the generator house was capable of supporting 16,000 pounds of additional equipment. To further conserve space, the drawworks power package, consisting of two model 8V-71, GM Detroit Diesel power units, was mounted on top of the drawworks transmission. The two Halliburton GM Detroit Diesel-powered well servicing pump sets were uniquely packaged to stack one on top of the other to save still more space.

The specially engineered Stewart & Stevenson power package met the severe performance requirements of a continuous 24-hour-a-day, 7 days-a-week operation for years of reliable service. In working with Gulf Offshore Company, Stewart & Stevenson helped make another valuable contribution to offshore technology.

Mud Pump Package

The first application of the GM Detroit Diesel 16V-149TI torque converter pump package for mud pump service was designed, built and delivered in six weeks to Pool Company Rig #17. The engine and pump package weighed approximately 71,000 pounds and could be broken down into two crane packages for offshore use and three units for land use. The pump was a National 9-P-100 Triplex Slush Pump and the maximum input horsepower was 1133. The Stewart & Stevenson modified Allison torque converter-reduction gear box with 1200rpm continuous output was coupled to a 22CB500 dual Fawick clutch. The pump produced 577 gallons per minute at a pump speed of 170. Other applications of this packaged design included the GM Detroit Diesel Allison 16V-92, 12V-149 and twin versions of each.

Helicopter Drilling Rig

A Helicopter Drilling Rig, developed for remote well sites in Venezuela, was engineered and designed into 4,000 pound modules in order for it to be transported by helicopter from one location to another. The new Helmerich and Payne Helicopter Rig, one of the most advanced in design, had the highest horsepower rating of any helicopter rig operating. Stewart & Stevenson supplied all of the power requirements for the light weight rig, which was equipped with five 12V-71T Detroit Diesel radiator cooled engines specially designed into 4,000 pound packages. Three of the engines were driving through GE 558 generators to an Allison TG602 transmission, which drove a specially designed Oilwell 350 PT pump. Two Detroit Diesel 12V-71T engines were driving 350kW DC Generator Sets for rig power.

Roads and navigable waterways were non-existent in the jungle drilling areas, so Tulsa-based Helmerich & Payne International Drilling Company solved all of the problems involved with six helicopter-transportable drilling rigs for operations in remote South American locations.

Air Drills Help Build Alaska Pipeline

Stewart & Stevenson Power Drills Air-Drilling Units were utilized in building the $6.37 billion Trans-Alaska Pipeline, approximately half of which was above ground. The above ground pipe was to rest on a cradle braced between two vertical supports, which were positioned about every 100 feet down the length of the above-ground portions. Some 80,000 holes, 24 inches in diameter and 25 to 35 feet deep, were required for the vertical supports. Drilling of the holes was accomplished over a 14-month period beginning June 1976.

Two Stewart & Stevenson customers drilled many of the 80,000 holes. Hughes Tool Company developed drilling equipment that featured a Detroit Diesel 12V-71T engine driving the air compressor, packaged by Stewart & Stevenson in Houston.

The emergency power generator set for the Texas Commerce Tower is lifted into place.

As part of an "anywhere, anytime" service tradition, George Witt and Floyd Cloer from Stewart & Stevenson's Houston branch joined Hughes Tool engineer Jack Smith the summer of 1976 for on-site start-up work and acceptance runs on the air rig. Special equipment was needed because of permafrost, a year-round condition along the pipeline route. Even in the middle of summer, the ground was frozen solid to within six feet of the surface.

Oil Industries Manufacturing & Engineering, Inc. of Odessa designed another piece of equipment for the air-drilling task. It was powered by an 8V-92 turbocharged Detroit Diesel. The Stewart & Stevenson Odessa branch worked with OIME on the systems. Six of the units were used by Parker Drilling Company of Tulsa to work on the pipeline, each reliably drilling four of the 24-inch holes per day.

S&S Drills Aid Skyscraper Construction

In 1979, four foundation drilling units designed and built by Stewart & Stevenson were used to drill foundation supports for the El Paso Tower, a project of Texas Commerce Bank and Gerald D. Hines Interests being erected on the Capital, Travis, Texas and Milam block in downtown Houston. Completed in 1980, the 75-story skyscraper was the tallest building outside New York or Chicago. Three of the drilling units had GM Detroit Diesel Model 4-53 engines driving through Allison 3331 transmissions. The fourth had a new model power system with an engine driving through an Allison MT-653 power shift transmission equipped with a push-button control.

Emergency power for the building, renamed Texas Commerce Tower upon completion, as well as for the Tower Annex Building, which housed computer systems, was also furnished by Stewart & Stevenson.

Chinese Order Frac Units

Stewart & Stevenson-built equipment for use in fracturing, cementing, and acidizing was being sought in far away places.

In 1975, the company was contacted by the People's Republic of China liaison office in Washington, D.C. regarding building 20 oilwell fracturing modules for China's oil industry. Stewart & Stevenson provided a bid based on the skeleton specifications the liaison office had provided, and "assumed that was the last we would hear from them," Slim Childress recalled in a 1999 interview.

But the liaison officer called and asked Stewart & Stevenson to send a representative to Beijing, China, where they were to deal with China National Technical Import Corporation of Beijing. Childress and Carsey Manning, Vice President of Major Contracts and Government Sales, flew to Beijing where they spent three weeks describing through an interpreter the equipment Stewart & Stevenson could provide.

The $7 million contract eventually agreed upon was for 12 pumper trucks, five manifold trucks and three blenders, all of which were truck mounted instead of trailer mounted for better maneuverability. The three blenders were hydraulic driven for precise control. The trucks were designed for 70-barrel-per-minute operations with 6,000 pounds of sand per minute. The tractors for each unit, blender, pumper and manifold truck, were all powered by GM Detroit Diesel engines and equipped with front tandem steering axles.

Remote operator panels, designed and manufactured by Stewart & Stevenson, provided for the control of all the equipment. The suitcase-size package permitted personnel to operate and monitor an entire fracturing operation at a distance of 100 feet. A headset communications system permitted constant verbal contact between key operators during operations.

After testing in Houston under the watchful eye of seven delegates from the Chinese firm in the spring of 1977, Stewart & Stevenson shipped the fleet of fracturing equipment, blenders and pump trucks to Beijing.

The $7 million fleet of oilfield equipment for shipment to Beijing undergoes manufacture.

ISC Advances Power Controls

Stewart & Stevenson formed International Switchboard Corporation (ISC) in 1972, primarily to supply the ever growing requirements of the powerhouse type switchgear for the Engine Division. A wholly owned subsidiary, ISC also became a major supplier of electrical control and switchgear for power generation and distribution. Its products were utilized by nuclear power plants, naval vessels, oil tankers, and in pipeline construction. Its SCR systems were used on drilling rigs.

One of the company's first projects was the main distribution switchboard of the new U.S. Navy DD963 Destroyer program. The switchboard successfully passed a series of shock and vibration tests to meet specifications.

In 1975, Stewart & Stevenson introduced a new and simplified turbine control system at the International Gas Turbine Conference. Made up of 17 printed circuit boards, the system was designed for starting, sequencing and controlling the Garrett-Airesearch Model GRP831-500 gas turbine engine.

The 500kW Garrett 831 turbine generator set.

SCR systems were custom-designed by ISC for the user; this one was for a petroleum company.

In 1977, ISC introduced a Silicon Controlled Rectifier (SCR) control system for both on- and offshore oil rigs, permitting Stewart & Stevenson to offer a single source supplier responsibility of complete AC/DC Diesel electric drilling systems for the oil and gas industry. The SCR system, custom designed for the end user, offered extreme flexibility, superior performance and packaging to aid in inspection and maintenance as well as wide application adaptabilities.

One of the oil rigs to be fully controlled by a new SCR system was the Pool Offshore Rig #453 working 30 miles offshore Louisiana. Pool's new modular design rig was designed to utilize every square foot of space. The SCR system was particularly appropriate under this concept.

During the oil field boom years of the late 1970s and the first couple of 1980, ISC did extremely well in SCR switchgear and oilfield controls markets, as well as building the switchgear and control setups under a subcontract with Ingalls Shipbuilding Company for the U.S. Navy Spruance Class Destroyers.

International Switchboard Corporation was sold to Equus Power Systems in June 1983.

New GPUs for the Airlines

By the end of the 1970s, Stewart & Stevenson aircraft ground support equipment could be found providing vital services at virtually every major airport terminal in the world. The company's units were supplying electrical power to planes while on the ground, providing compressed air to start jet engines, and pushing or towing aircraft into or out of tight places.

Growth in the airline support business continued with new orders for its pneumatic ground power equipment for servicing the L1011 and Boeing 747. Each truck mounted unit was capable of delivering up to 2709 pounds of air per minute at 41 psig. Two separate screw type air compressors, powered by two Diesel engines, were housed on a conventional truck chassis. The pneumatic power units were designed primarily for aircraft engine starting, continuous operation of aircraft environmental (air cycle) equipment, cabin pressurization testing and operation of various aircraft system components. They were capable of servicing the Boeing 707, 727, 737, and 747; Douglas DC-8 (all series), Douglas DC-9 and 10; Lockheed L1011 and Convair 880.

Truck mounted air start unit.

New Tow Tractor

In 1977, Stewart & Stevenson introduced the Gater, a sensible alternative to the expensive, monster aircraft tow tractors then on the market. It featured less initial cost, less fuel consumption, less noise, less pollution and less downtime. Using the Detroit Diesel Allison Model 3-53N Diesel engine, the tow tractor delivered up to 14,000 pounds of tractive force through a smooth Allison AT 540 automatic transmission with four forward and one reverse gear. The 12-foot long, 80-inch wide Gater weighed 28,000 pounds with maximum equipment.

More Powerful Gater

In 1978, Stewart & Stevenson introduced a more powerful gate tractor capable of pulling any size aircraft. A "big brother" to the previous Gater aircraft gate tractor, the "Greater Gater" was built to easily handle the largest aircraft. At 15 feet, 2 inches in length, the newest Gater was still smaller than most other gate tractors, but it had a tractive effort of 40,000 pounds. The tractor was designed to move DC10s, 747s, and L-1011s. The unit weighed 60,000 pounds and optional equipment included a 400 Hz electrical ground power unit, operator cab, radio intercom, air compressor and storage, fifth wheel attachment, customer specified lighting, enclosed 2-man cab and radio intercom cabling. The Gater was equipped with a dependable GM DD 4-53 Series engine driving through an Allison TT-2421-1 transmission with four forward and two reverse speeds. The Gater had an all steel welded frame

Tow tractor with rear mounted ground power unit.

The GT-32.

Stewart & Stevenson Ground Power Units ready for delivery.

and ran on standard tires. Tractor parts were interchangeable with many of Stewart & Stevenson's ground power and air start units.

Plane Mate

"Plane Mate," a new airport shuttle system that shuttled passengers between the terminal and the plane parked some distance away in a non-congested area of the airport, was powered by Stewart & Stevenson Diesel propulsion units and the Diesel generator packages. The Port of New York acquired eight of the shuttles, Dulles International Airport had twelve, Eastern Airlines had three, and the Montreal Department of Transportation had three.

Air conditioning unit mounted under jetway.

Left: Ground support for Air Force One, built by Stewart & Stevenson.

Gas Turbines: A Company Turning Point

'There's a Turbine in Your Future'

Stewart & Stevenson began developing gas turbine powered generating units in the early 1970s. The first, a 2500kW trailer-mounted standby power unit, designed to provide a large block of portable power in emergencies, was completed in 1971. The company had sold three or four gas turbine packages when the Navy contract for Spruance Class Destroyers came up for bid.

Designed to be the first ever all gas turbine powered ship, the Spruance Class Destroyers were conceived to have a propulsion system powered with four General Electric LM 2000 gas turbine engines (the marine version of the company's jet aircraft engine) and the ship's service electrical generators were to be powered with 2000kW gas turbine units.

Litton Industries had the contract to build the ships at its Pascagoula, Mississippi plant. Carsey Manning, manager of Government Sales at that time, made a deal with Allison's management to sell Stewart & Stevenson the Allison gas turbine engines, which Stewart & Stevenson packaged for the gas turbine sets.

A $42 Million Navy Contract

The $42 million contract, awarded in 1971, called for 90 gas turbine generator sets for the Navy's DD963 superdestroyers, three 2000kW generator sets for each of the 30 destroyers to be built. At that time, it was the largest subcontract ever awarded to a firm qualifying under the U.S. Government Small Business Administration.

Orchestrated by Carsey Manning, that contract was another turning point for Stewart & Stevenson.

By 1972 Stewart & Stevenson had customized the Allison 501K gas turbine for industrial, marine and generator applications. The Allison turbine opened new horizons for industrial, oilfield and marine uses requiring 3,000 to 5,000 horsepower in single gas turbine packages.

In 1973 Stewart & Stevenson introduced a unique, compact "stacked" turbine, with models ranging from 1500kW to 6000kW and above. It also was powered by the Allison 501K single shaft gas turbine.

In 1974, Stewart & Stevenson began packaging Garrett industrial gas turbine engines. The Model 831, with a horsepower range up to 800bhp, filled the gap between the Allison 501K and smaller units. Major markets for the Garrett 831 included offshore oil production platforms, emergency generator sets, pumping units and marine applications.

Below: Allison gas turbine engine powered generator.

Turbine Era Progresses 1980 -1990

Multi-Million Dollar Turbine Contract Jump Starts New Decade

The year 1980 began with Stewart & Stevenson signing a multi-million dollar contract to furnish generator sets for an offshore production platform in the Indian Ocean off the West Coast of India. The contract from the Oil and Natural Gas Commission (ONGC) of India called for three 15 megawatt gas turbine generator sets, each powered by the General Electric LM2500 gas turbine engine. It was a risky venture into a new, relatively unknown field, but Stewart & Stevenson confidently took on the project.

The three modules, two for prime power and one standby unit for a single offshore platform located in the Bombay High North producing field, were shipped and put into operation in 1982. The giant production platform, which handled almost 20 different producing wells, processed and dried natural gas, separated oil from gas, separated gas from water, and sent the oil and gas by pipeline on to land. The platform had quarters and full facilities for more than 100 employees.

India was relatively new to the offshore oil and gas exploration industry, but was rapidly expanding its drilling and production capabilities in the Indian Ocean. Stewart & Stevenson had previously supplied ONGC with small 30kW dual fuel generator sets for individual production platforms.

The GE LM2500 gas turbine, rated at 30,000 horsepower, had been proven with successful applications for power generation systems, gas compressors and ship propulsion systems. The single-lift factory packaged concept pioneered by Stewart & Stevenson for the ONGC project marked the beginning of the company's expanded turbine era with GE aeroderivative gas turbines.

The Rise Before the Fall

Growth in Stewart & Stevenson's Engine Operations resulted in ever increasing sales: a 28 percent increase in 1979 to a high of $268 million; a 32 percent increase in 1980 to a new high of $354 million; and a 31 percent increase in 1981 to a record $463 million.

Then, the oilfield recession of 1982 hit, resulting in the company's first loss in more than 20 years. Sales decreased 20 percent to $371 million, with all industry segments reporting declines. Only government sales increased, up 80 percent over the previous year, due largely to the United States' renewed commitment to a superior Navy and Stewart & Stevenson's concentrated effort to increase penetration of this market.

Particularly hard hit were the energy and marine markets in the company's engine and electrical equipment operations. The world had moved almost overnight from an oil shortage to an oil glut, and the sharp falloff in oil field drilling activity resulted in numerous cancellations of existing orders and an abrupt halt in new orders. Even Diesel parts and service sales, which historically were immune to recessions, reported significant sales declines. Instead of ordering parts from Stewart & Stevenson, companies with rigs needing parts took them from rigs they owned but no longer had in operation. The only glimmer of a silver lining to this cloud was that the banks, which repossessed rigs that drillers couldn't pay for, had nowhere to store them nor any means to service them and turned to Stewart & Stevenson to provide this service.

Sales continued to decline in 1983, down 36 percent to $237 million. The company recorded a loss of approximately $1.5 million. The major factor contributing to the loss was the annual carrying cost of approximately $8 million for the oilfield equipment — one jack-up drilling rig, two posted barge rigs, two anchor handling tugs and one offshore supply boat — for which contracts were canceled. Stewart & Stevenson had to write off $23 million of non-recoverable costs due to customer cancellations of contracts for offshore drilling rigs and related vessels.

When the Times Get Tough...

As a result of underutilized manufacturing facilities, Stewart & Stevenson immediately reduced employment levels by almost 50 percent. From a peak level of 3500, the number of employees by 1983 had dropped to 1700. Salaries of remaining employees were frozen, and the company consolidated operations, closing certain warehouses and manufacturing areas. Nonessential overhead expenses were reduced and planned capital expenditures were delayed. In June 1983, the International Switchboard subsidiary was sold.

Keeping a long-term perspective, Stewart & Stevenson continued its strong research and development efforts and redirected marketing efforts in new areas. The company would emerge a leaner, stronger company with reduced costs and improved products.

Left Page: The giant ONGC production platform in the Indian Ocean.

Building on the Positives

While Stewart & Stevenson was the world's largest distributor of Diesel engines, the company's success had been built on using those engines as prime mover to build anything that required a Diesel engine as the prime mover. The company had a long history of innovation.

With a standard factory engine, the company had built in heavier duty cooling and lubrication, and increased fuel injection to create a durable marine propulsion unit with more horsepower.

Stewart & Stevenson had converted a Detroit Diesel engine to run on a mixture of 85 percent natural gas and 15 percent Diesel fuel, and engineered it to operate in a vertical position. On offshore production platforms, the vertical, dual-fuel units took up less space and ran on gas produced from the well with a small charge of Diesel fuel for ignition.

Stewart & Stevenson had applied extensive engineering and sophisticated electronics to "marry" the engine and generator and to assure precise operation of its generator systems. And that was just the starting point. The company further customized them to operate at minus 60 degrees Fahrenheit in Alaska, or at 120 degrees Fahrenheit in the Saudi desert, and to withstand the salt spray and punishing environment of the North Sea.

1984: Headed Back Up

Building on these positives and a "back to the basics" sales approach, Stewart & Stevenson emerged a much stronger company in 1984. Sales rose from $237 million to $257 million, with an after-tax profit of $3.4 million.

Stewart & Stevenson generators found industrial use as standby power units for a Dallas hospital and for the Riyadh International Airport. They provided electricity for a major dam and hydroelectric construction project in South America. They powered road building and earth moving equipment, pumps and compressors of all kinds, and farm irrigation systems.

The company completely redesigned most of its generator line with more economical, simplified controls.

The company also built on its vast military experience, which began with repowered tanks and generator sets for the Allies in World War II and continued in the 1950s and 1960s with generator sets to power communications systems, Nike missile launchers, sophisticated radar installations, aircraft ground support systems, and the nation's space program. The company's diversified activities for the military included everything from air-transportable 750kW mobile generator sets to tow tractors for the Air Force's enormous C-5B transport aircraft.

Stewart & Stevenson's 1,000kW Detroit Diesel-powered generator sets were providing the ship service electricity for FFG-7 "Oliver Hazard Perry" class guided missile frigates. The company's tow tractors were handling planes on U.S. Naval aircraft carriers, and its Diesel engines were found on carrier generator sets, spotting dollies, fire trucks, hydraulic test stands and air conditioners.

By 1984, Stewart & Stevenson had delivered more than 120 Air Force MB-2 tow tractors, which had a drawbar pull of 27,000 pounds, for use with KC-135 and B-52 aircraft.

Projects for the military included engine test stands, mobile and stationary pumps, specialty vehicles, repowered vehicle propulsion systems, marine propulsion systems, new aircraft ground support services, cogeneration units and equipment service-life extension programs.

A Stewart & Stevenson generator provides standby power for the Riyadh International Airport.

Stewart & Stevenson's 1000kW Detroit Diesel-powered generator sets provided ship service electricity for FFG-7 "Oliver Hazard Perry" class guided missile frigates.

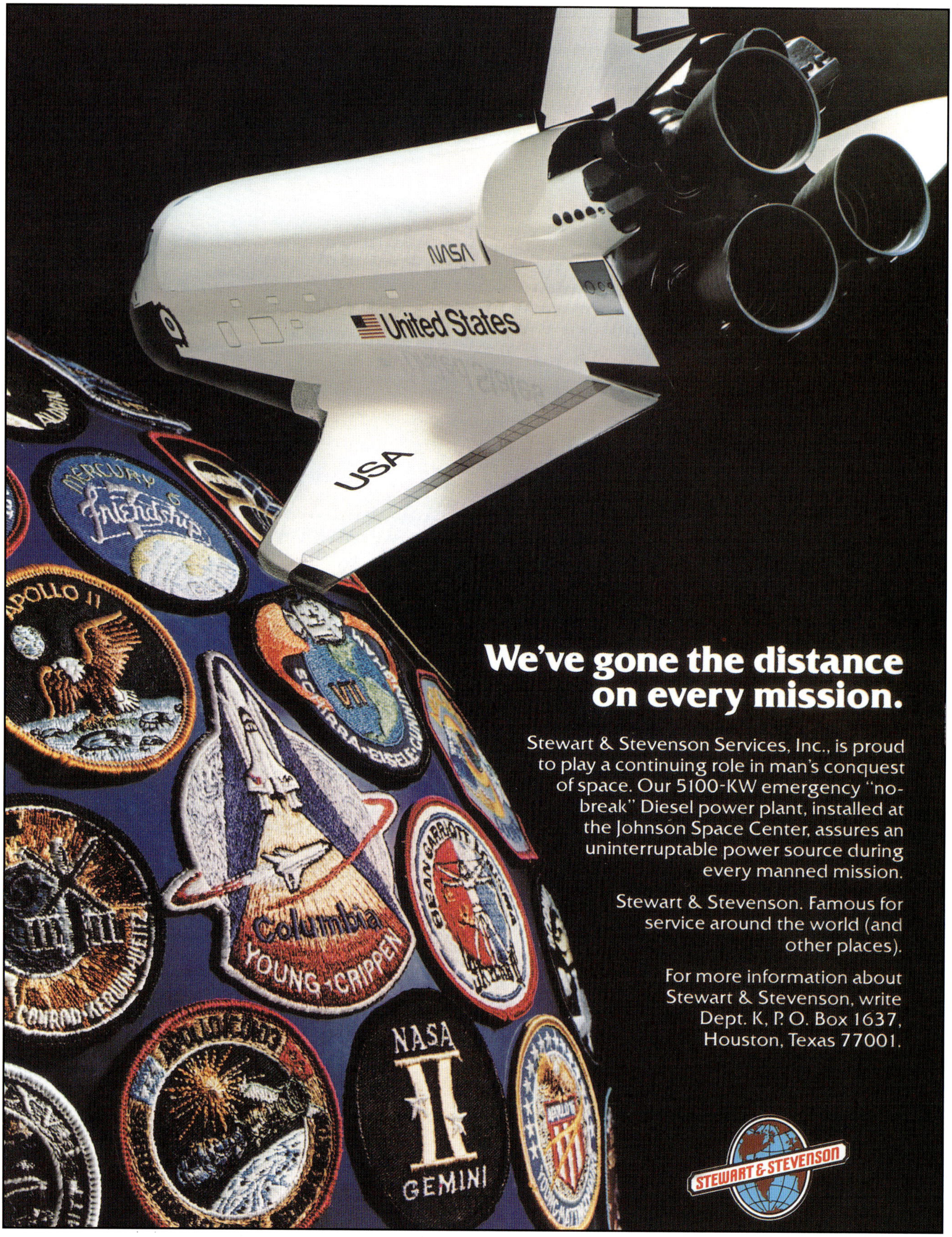

A Stewart & Stevenson advertisement that ran in the early 1980s.

The Air Force MB-2 tow tractor, designed for use with KC-135 and B-52 aircraft.

The EGT-50 electric powered tow tractor.

In addition to ongoing technical services for the Department of Defense, Stewart & Stevenson's government business included equipment for the U.S. Coast Guard, the Army Corps of Engineers, NASA, the Department of Transportation, and the Department of Energy.

Howard T. Tellepsen, Houston's fire boat, powered by the 12V-71TI.

Stewart & Stevenson's mid-sized GT-50 tow tractor.

Stewart & Stevenson's aircraft ground support equipment successes included the smaller, more maneuverable unitized tow tractors, in Diesel and battery/electric versions, and more efficient air-start units. The tow tractor's body and frame were a single steel weldment, with pockets built in for proper distribution of ballast. The company's newest version was the mid-sized GT-50 tow tractor. Its newest air-start model was the TMSS/TRSS-120. Both tow tractors and air start units were in use by airlines around the world.

Famous for engineering specialized marine propulsion systems and building complete boats around them, Stewart & Stevenson introduced a new highest horsepower line of marine Diesel engines in 1984 for the pleasure craft and offshore patrol boat market.

Capitalizing on technology developed for the U.S. Navy, the company provided gas turbine propulsion packages for European and American boat builders. The new version of the original "Sweet Sixteen" was the 16V-92MTA. The company delivered more than

This 16V-149-TI, dubbed the "Mean Machine," was designed for the large yacht market.

A $15 million well-servicing equipment order for an overseas company is loaded for shipping.

100 engines in 1984 to police and patrol boat builders from Florida to Hong Kong to Venezuela.

Stewart & Stevenson also continued to supply package boats, from 45 to 200 feet in length in three basic types: inland pushboats to move barges along the nation's inland and intercoastal waterways; crewboats to transport crew and high priority cargo to offshore platforms; and offshore vessels to carry heavy cargo to production platforms and to position drilling rigs.

In the petroleum market, Stewart & Stevenson in 1984 was able to sell or lease a significant portion of its inventory resulting from default by customers when the oil field crashed. The company continued to specialize in three facets of the petroleum marketplace: drilling and production, well servicing and well stimulation.

The LM2500 gas turbine offered new possibilities in offshore applications. In a cogeneration configuration, these systems produced both the electrical power and steam for heat for living quarters on multiple interconnected platforms, as in the ONGC application.

For the well servicing marketplace, Stewart & Stevenson provided both mobile and electrical power for workover rigs. But the company's most profitable sales were in well stimulation: frac units that used sand, water and chemicals under high pressure to fracture oil bearing formations; blenders that mixed compounds for frac units; nitrogen units that used cold nitrogen gas to increase yield; and acidizers that cleaned out the "gunk" to prevent it clogging a formation after it had produced for a time. The company completed development of a new frac unit in 1984, which at 2000 horsepower was the largest in the world at that time.

Highlighting 1984 sales was the delivery of $15 million in well-servicing equipment to the People's Republic of China, and an additional $15 million in orders from the Chinese scheduled for 1985 delivery.

In February 1984, Stewart & Stevenson reached an agreement with MWM, a major German Diesel manufacturer, to handle and package their line of large-bore and heavy-fuel-burning engines. In addition to adding heavy-fuel-burning capabilities in the company's 2000 to 4000 horsepower range, the large-bore MWM Diesel engines extended Stewart & Stevenson's horsepower range from 4000 horsepower to 8800 horsepower.

New World Headquarters

In the spring of 1984, Stewart & Stevenson moved into its new world headquarters at 2707 North Loop West in Houston. The company's executive offices and many of its departmental groups moved into the Stewart & Stevenson Tower, constructed and owned by Mischer Development Corporation.

On the wall of the new World Headquarters Board Room was the 1919 newspaper article with a photo of the company's then new headquarters building at 1713 Preston. The last paragraph asked the question: "Will a house resting upon such a foundation endure?"

Now in its 82nd year, Stewart & Stevenson had indeed endured and continued to build on its proven foundation of goodwill and honest and intelligent service. The new world headquarters reflected its status as a globally growing business.

One year after Walter Mischer, Texas Governor John B. Connally, and C. Jim Stewart II (left) broke ground for the Stewart & Stevenson Tower, the company moved into its new world headquarters at 2707 North Loop West (below).

Thomas H. Langham III, President (1985-1988).

Thomas H. Langham III, President

Thomas H. Langham III, son of Aileen Stewart Langham, daughter of founder C. Jim Stewart, became President in 1985. He joined the company in 1949 as a salesman in the Waco, Texas territory. Tom had 36 years of broad experience with the company, having served in many areas of the company's operations, most recently as executive vice president. Under Langham's direction, Stewart & Stevenson grew and branched into new areas. During his tenure, the Gas Turbine Division was formed and John Deere and Hyster equipment sales activity increased, as did demand for other Diesel products.

Throughout his career, Langham spent time in the plant, with the sales force, and with customers. Company employees were inspired by his willingness to personally undertake any challenge, his insistence that the company do what was right, not just what was required, and his good humor in the face of adversity. On January 25, 1988, Langham passed away after a courageous battle with cancer.

Cogeneration: A New Frontier

Long a stepchild among prime movers, the aeroderivative gas turbine had been refined and expanded to become a viable contender, with excellent fuel consumption. Among its many advantages was its exceptionally low weight-to-horsepower ratio — up to 100 times less weight than that of a Diesel engine of the same power. Waste heat generated by the gas turbine, utilized by new cogeneration technology, provided even more efficiency. The thermal energy produced could be used to fire steam boilers, provide air conditioning or operate a multitude of other machinery. Since legislation passed in the late 1970s had deregulated the utility markets, cogeneration from a turbine could also be sold to many private investors.

Since the late 1960s Stewart & Stevenson had packaged gas turbines for prime power generation, pump drives, gas compression drives, marine propulsion and cogeneration. Experience gained with Garrett and Allison gas turbines served as a springboard for packaging large multi-megawatt electrical generator and cogenerating plants.

Two Allison 570 gas turbine powered units provided the Boeing manufacturing facility with its electrical, heating and cooling needs.

Garrett cogeneration unit.

Following the ONGC project, Stewart & Stevenson in 1981 designed and built two cogeneration units for Boeing's manufacturing facility in Washington. Two Allison 570 gas turbine powered units provided the facility with its electrical, heating and cooling needs. In 1983, the company designed and packaged a self-contained single lift 21.5 megawatt cogeneration unit utilizing a LM2500 GE gas turbine for a Pacific Resources Inc. oil refinery in Hawaii.

Stewart & Stevenson went after the strong cogeneration market in both the Diesel and gas turbine fields, signing contracts for seven cogeneration units in 1984. In 1985, the company delivered two units each to Shell Oil Company, Conoco, and Hawker Siddley, and a single unit to United Airlines. The two Shell cogeneration units produced steam for enhanced oil recovery, and the company ordered a third. Conoco's dual fuel units were for use at their Milne Point project on the north slope of Alaska.

At the United Airlines maintenance facility in San Francisco, heat from the Stewart & Stevenson cogen-

Two GE LM2500 gas turbine powered cogeneration units, owned and operated by Sunlaw Energy, furnish power for cold storage plants in the Los Angeles area.

The dual fuel units for Conoco's Milne Point project on Alaska's North Slope pass the San Jacinto Monument as they leave the Port of Houston. The barges on which they were loaded were towed by tugs through the Panama Canal and then up the West Coast to Alaska near Anchorage. The units were then transferred to another barge for the final leg to the North Slope.

The SoHio 501KB cogeneration unit.

eration unit provided steam both for heating the entire facility and for cleaning aircraft. In addition, 20 megawatts of simultaneously-generated electrical power was sold to the local utility company, helping to offset United Airlines' cost of owning and operating the unit.

As was true for the Diesel engines designed and engineered by Stewart & Stevenson, the company's modular, factory tested gas turbine systems usually paid for themselves in less than three years.

Units for cogeneration generated sales in 1985 in excess of $30 million, and interest in the company's 25 megawatt gas turbine package increased. Stewart & Stevenson began work on several cogeneration packages for enhanced oil recovery projects near Bakersfield, California, a unit for the University of California at Berkeley, and a package designed for the city of Wellington, Kansas.

In 1985 Stewart & Stevenson formed California Cogen, which evolved into Stewart & Stevenson Operations, Inc. (SSOI), a wholly owned subsidiary under the direction of Don Wallin. A complete operation and maintenance business for the power producing industry, SSOI's operations extended across the continental U.S. and South America and expanded into worldwide operations.

In 1986, Stewart & Stevenson formed the Gas Turbine Products Division to handle the increase in business in gas turbine-powered equipment.

Rick Stewart, C. Jim II Recount Gas Turbine Growth

As manager of Harrisburg Manufacturing, Rick Stewart became knowledgeable about gas turbines when Stewart & Stevenson performed the military contract for the Spruance Class Navy Destroyers.

"At 10 and 11 o'clock at night we would be on the test line to stay on schedule with the Navy," he said.

Stewart called the company's early commercial turbine manufacturing "a disaster because of the tough competition. The price was low so the oilfield companies would buy them, but each was custom designed by the customer's engineers. We would build them, then 30 of their engineers would critique what we built. With competitive pricing, we lost a fortune on each one. In 1983, I said we needed to get out of the gas turbine business for oil companies.

"About that same time, we were finishing our first ONGC LM. It was so huge it scared us. We needed a larger building, and it was a real challenge to test them. On Friday afternoon, we would call all the hourly workers in and hand out $100 bills to them, tell them to take their wives out for dinner and give them the news that they had to work all weekend. We had to deliver the ONGC LMs on time, and we did. The units are still running in Bombay," Stewart said.

Then cogeneration caught on. Stewart & Stevenson built five, all delivered by 1985.

"Those five units kicked us off in the cogeneration business," Stewart said, "and since we built all the same and at the same time, we made fair money on them."

"But Stewart & Stevenson could not continue to build the huge gas turbines at the Harrisburg plant," Chairman of the Board C. Jim Stewart II said. "Our policy was to test everything before delivering it to the customer, and one of the problems we found in starting up the huge turbines for testing at the Harrisburg plant, which is kind of funny, was that every pilot light in every house for a mile around the plant would go out. The gas company and city were in an uproar about us running these turbines within city limits. They were like a 747 aircraft engine starting up next door."

So C. Jim II told his son Rick to determine what was needed in the way of space and facilities and either come up with land and a building design or find a building to lease. Because it was uncertain if the market would stay, Rick decided to lease a building he found "in the middle of the weeds and marsh" at Jacintoport.

Rick Stewart decided that he wasn't willing just to be a manufacturer of gas turbine packages. His goal was to build the turbine business into a major profit center for the company. Weldon Janek selected engineers and manufacturing personnel to move to Jacintoport, and salesmen Mark Axford, Jay Manning, Bill Fly and Jim McKastle joined the force. A total of 34 people moved with the Turbine Division to the Jacintoport facility.

The cogeneration market grew and the Turbine Division grew with it. When the U.S. market slowed, it was offset with international business. Stewart & Stevenson shipped cogeneration units worldwide, to Russia, the Amazon, Tanzania, Africa, all over Europe, China, Malaysia, Indonesia, and the slopes of Alaska.

"I can't think of any place where we didn't have a presence or were not calling on people," Rick Stewart said. "We were the smallest company in the business, competing with GE, Westinghouse, Siemens, and all the multinational companies, but everybody in the industry knew us well. We were just a nice bunch of Houston oilfield boys in the gas turbine business, and we never got a big head on making a deal. Customers respected us for our aggressiveness, the quality of our product, and our service responsiveness."

Rick Stewart was successful in attaining his goal. In 1986, the Gas Turbine Division did $20 million in business. By 1995, it had grown to $650 million. With Service and Maintenance added to Original Equipment Manufacturing and Sales, the division grew from one to eight buildings with quadruple the space and from 34 to more than 2,000 employees.

Then, an industry-wide slowdown in both the domestic and international markets resulted in more competitive pricing and lower profits. With so many companies competing for the same business, it was inevitable that consolidation would begin.

"There was no longer room for two sets of profits in the gas turbine chain," Stewart said. "GE, which was selling gas turbines to Stewart & Stevenson, had to move away from the two-step distribution business."

The Vice Chairman of General Electric called Rick Stewart and invited him and his father to lunch. The end result was "they made us an offer we couldn't turn down," C. Jim II said, "despite the fact that we hated to give up the business." The offer included having Rick Stewart stay with the gas turbine business for GE.

"The great thing," Rick Stewart said, "is that Stewart & Stevenson had built the gas turbine business to have such a great reputation in the industry that GE wanted to buy it. GE could have just eliminated Stewart & Stevenson's distributorship franchise.

"It was an honor to Stewart & Stevenson and its whole philosophy of doing business, our entrepreneurship, and our great people, that GE wanted to buy the gas turbine business," he concluded.

The original Jacintoport Turbine Division facility in 1986, above, had expanded into eight buildings by the time it was sold in 1997, below.

Restructuring for Growth and Earning Potential

Early in 1986 Stewart & Stevenson began realigning the management and organization of the company into business segments that would provide growth and earnings for the future.

Electrical Equipment Operations, which consisted of the subsidiaries International Switchboard Company and International Electric Corporation (IEC), ceased to be financially significant with the 1986 closure of IEC's generator manufacturing plant and licensing Baylor Company to build the IEC designs, putting an end to the continuing losses generated by this line of business.

The remaining lines of business were restructured into two divisions: Engineered Power Systems and Distribution.

Engineered Power Systems

The Engineered Power Systems Division manufactured gas turbine-driven equipment, military products, aircraft ground support equipment and other highly-specialized power applications.

Turbine-driven equipment accounted for more than one-third of the division's sales in 1986, as the company consolidated its leadership in the manufacture of cogeneration units. In anticipation of further growth, Stewart & Stevenson relocated commercial turbine equipment manufacturing and support operations to a new facility at Jacintoport.

In 1986, Stewart & Stevenson received a grant from the Gas Research Institute to develop a 100kW reverse-flow, ebullient-cooled natural gas cogeneration unit.

The company received a $12.7 million contract for nine 2,500kW Allison gas turbine generator sets for three Aegis Cruisers, and a $9 million contract to supply Diesel generator sets for the U.S. Army's new Quiet Reliable Generator Program. Contract options called for up to $50 million in equipment over four years.

Under previous government contracts, the company delivered eighteen 2,500kW Allison gas turbine generator sets for six Aegis Cruisers, sold and delivered eight 1,000kW Diesel-powered generator sets for installation on two guided missile frigates being built for the U.S. Navy and Australian Navy; sold and delivered two 2,000kW Diesel-powered generator sets for the U.S. Navy's new LHD helicopter deployment ship, and delivered 31 tow tractors to the U.S. Air Force as part of a $5 million contract.

Stewart & Stevenson introduced a new GT-1628 tow tractor and began deliveries of the new BT-45 baggage tractor for airline ground support.

In the marine market, Stewart & Stevenson received a contract for 68 DIESELDRIVE® 8V92TA 736 horsepower engines from an East Coast boat builder, with options for up to 100 more. The company also supplied engines for a prototype 65-foot Sport Fisherman, leading to sales for 12 more; received a purchase order

This turbine generator set was destined for the U.S. Navy's *Ticonderoga*, a CG-47 Class Guided Missile Cruiser. The generator is powered by an Allison 501 gas turbine driving a 2500kW generator.

The Aga Khan's yacht was powered by the Allison 570 turbine.

Allison 570 turbines also were supplied to the Swedish Navy.

The BT-45 baggage tractor for airline ground support.

for 60 engines for military patrol boats, and provided the propulsion engines that carried the General's Titan motor yacht to victory in offshore racing's most famous event, the 362.3-mile Miami to Nassau to Miami race.

The petroleum market continued to worsen in 1986, as even the Chinese were now taking a "wait and see" attitude. Stewart & Stevenson began a Rig Maintenance Division to store rigs owned by non oil-related companies, such as banks and leasing companies, which had no in-house maintenance capabilities. On the positive side, as the market continued to worsen, much of the company's competition was eliminated, leaving Stewart & Stevenson one of the few companies available to build and supply rig power and rigs for the petroleum market when it did pick up.

Distribution Division

The Distribution Division consisted of franchised distribution of Diesel engines, transmissions and spare parts; rebuilding Diesel engines and automatic transmissions; sale of material handling equipment, truck refrigeration units and other industrial equipment; and providing repair and maintenance services.

Principal distribution territories included Texas, Colorado, New Mexico and parts of Wyoming and Nebraska.

The 1986 market for the division was characterized by the worst regional economy ever faced by Stewart & Stevenson. Falling commodity prices for oil and natural

Stewart & Stevenson delivered six 2300kW dual fuel emergency generator sets to provide backup power at the Consolidated Space Operation Center in Colorado. The generator sets are powered by Mirrless Blackstone engine driving International Electric generators.

gas affected everyone engaged in commercial or financial activity in the Southwestern United States.

Stewart & Stevenson took advantage of the adverse environment to expand the types of products handled by the Distribution Division, add new territories and improve the efficiency of its facilities.

Acquisitions included Sandia Detroit Diesel in Albuquerque, the New Mexico distribution territory for Detroit Diesel engines and spare parts; Amarillo Diesel, which added a full-service branch to West Texas and gave the company a presence in the Texas Panhandle; the North Texas distribution territory for Hyster Company material handling equipment; and the Casper, Wyoming dealership for Deere & Co. industrial equipment.

Throughout Texas, Stewart & Stevenson consolidated Diesel equipment and material handling equipment branch locations in the same city.

Stewart & Stevenson signed a distributorship agreement with the American affiliate of Germany's M.A.N. Corporation, the world's oldest manufacturer of Diesel engines, to market M.A.N. cogeneration modules in all areas of the U.S. west of the Mississippi River, plus Louisiana, Mississippi, Alabama, Alaska, Hawaii and the Caribbean.

The division also commenced production of rebuilt torque converters for General Motors Corporation.

Parts, Service – Anytime, Anywhere

Stewart & Stevenson's long history of providing its customers with around the clock, around the globe parts and service continued to play a key role in the company's success.

Three reports in the company's 1986 annual report amply illustrate the company's "Parts & Service... Anytime — Anywhere"™ philosophy.

The *Kinkaid* – Somewhere in the Pacific

Jim Vickers, Stewart & Stevenson's service representative in San Diego, got back into the office at 10 a.m. only to find out the Navy wanted him in the air as soon as possible. They weren't telling the details. That would come later — on the way to the Seychelles Islands.

Jim made it as far as London uneventfully. But as his DC-10 taxied out to leave Heathrow Airport, he was summoned to the cockpit.

"I don't know who you are," said the pilot, "But you're supposed to deplane and reroute." As the plane pulled to the side of the taxiway and a pickup sped to get him, the pilot offered condolences. "What do you want us to do with your baggage," he asked, "after it gets to Africa?"

Jim had no idea he would spend the next sixteen days in the clothes on his back.

The ship he had been trying to connect with had moved. Jim was to go to Hong Kong via Germany and Saudi Arabia, as the first step in reaching the new location, wherever it was.

"After four hours of trying to get a hotel in Hong Kong, the place I finally got was a flea bag," he says. "It was late, I hadn't eaten and there were no restaurants. Right outside the hotel, they were killing chickens in the gutter, and there was a coffee table, two chairs, and a kerosene lamp. So I got a midnight snack."

The next day, Jim flew to Manila in the Philippines and was met by a driver who took him to Clark Air Force Base. Once there, nobody seemed to know why he had come. After initially refusing to let him use the phone (in spite of DOD clearance and government orders, he was still a civilian), the Navy liaison relented, and they ultimately discovered that he was to proceed to Diego Garcia, a controlled-access island in the Indian Ocean.

After a 2 a.m., eight-hour military flight to that island, Jim was transferred to an S-3 jet for the five-hour flight to the deck of the USS *Enterprise*. "We were strapped in, straps across our shoulders, down across our waists, up between our legs like in a fighter plane," Jim says, "because just before they land, they flip around in the sky and pull about four Gs. You feel like the skin on your body is departing you."

On the *Enterprise*, Jim escaped the care of the petty officer, who planned to keep him in the holding area for a couple of days, and found the squadron commander's office. The two of them went through the message traffic to discover that the trouble concerned one of the three generators on the destroyer *Kinkaid*. Jim scavenged parts from some P-2 jet engines in the carrier's maintenance shop ("I was still wearing the same clothes, and crawling up the exhaust of this engine to get to the heat shield cone."), and then proceeded via three helicopter-to-ship hops to the *Kinkaid*.

Although the repair made the engine serviceable for emergency use, Jim decided it should be replaced. Another story: locating the engine (Jim knows Navy engine inventory by heart); building rigging equipment to move the crated engine from the tender ship to the *Kinkaid*; and working 17 hours around the clock to complete the installation.

"Sleep?" Jim says. "There was no time to keep your eyes shut."

Jim remained on the tender after the job was completed, and the next complication arose because that ship had no helicopter landing pad. Finally a motor launch took him to a supply ship, where he reached the deck by climbing hand-over-hand up a cargo net hanging loosely over the side as the ship heaved in the sea. From there by helicopter to Oman, and from there back to Diego Garcia, Manila, Taiwan, Tokyo, Seattle, Los Angeles, and home.

"You don't need a pat on the shoulder, you just go and do your job," says Jim. He laughs about his adventures and obviously enjoys his work. "As long as everything runs right, you've done your best, and that's the way I see it. I haven't had a failure yet that I haven't gone out and fixed."

To War-torn Central America

"The Navy SEALs told me it was U.S. Embassy rules that non-military personnel on government business were supposed to have two armed guards at all times. Then they said that there were only six of them, and they all had a job to do, so I would have to get by with a M-16, an Uzi sub-machine gun, and a .45 pistol."

Bret Nichols, Stewart & Stevenson service representative in Houston, talked about his adventures servicing an S&S engine aboard a military patrol boat in one of the war-torn countries of Central America.

Arriving in the country at the request of SwiftShips,

Inc. of Morgan City, Louisiana, the builder of the patrol boat needing service, Bret was met at the airport by an armored car with four guards carrying machine guns. He spent his first night under guard at a "safe" hotel. The next day, after an hour's ride on a military aircraft, he arrived at the country's naval base and took up quarters at the U.S. Navy compound nearby, where he was issued the above-mentioned arsenal and a flak jacket.

"We had to drive through this little town, about 300 people, to get to the boat I was working on," Brett recalled. "Now one of the common things down there is they say people with beards are the guerrillas and the people without beards are the good guys. Well, I happen to have a beard..."

In spite of his looks, Bret avoided being shot at in what he describes as a very civilized war. "You'd hear the gunfire start about six in the morning," he explained, "and it goes 'til noon, and then quits. Everybody takes a siesta, and the gunfire starts back up about 2:30 and goes 'til dark."

To the accompaniment of this near constant machine gun fire, Bret fixed the patrol boat's engine, stopping only to head for a bunker when the bullets seemed to be getting too close. Once the engine was working properly, Bret proceeded to repair everything else on board the vessel that needed it. "I fixed the toilet system, repaired the radar, worked on the electric bilge pumps, a little bit of everything," he said. "SwiftShips is one of our customers and they depend on us to work on anything."

Calvin LeLeux, vice president of SwiftShips, concurred. "We have worked with Stewart & Stevenson ever since our company began in 1969," he said, "and the reason is because Stewart & Stevenson has a more well-rounded and international posture than most of the other engine distributors and vendors. They know how to solve problems on an international scale. And they help out with other items that need fixing. It's hard to send an expert in every field to service a vessel in a hostile environment."

SwiftShips builds specialized medium-to-high-speed vessels, ranging from 30 to 200 feet in length, which it sells primarily to foreign militaries. Along with vessels, SwiftShips also offers service contracts which Stewart & Stevenson personnel help to fulfill.

Bret Nichols says he enjoys the work "I've been almost everywhere there's a conflict," he said. "It's just the way it is."

A Weekend Service Marathon

It was Friday afternoon when the work crew at Houston's Metropolitan Transit Authority discovered that the gear assemblies for the power steering pumps on three engines that Stewart & Stevenson delivered that morning were in the wrong place. An emergency — the engines had to be ready for installation first thing Monday morning.

The engine order sent in by Metro had not specified the location of the gear assembly, and most buses were designed differently than these three. To reconfigure the engines, "we had to pull the whole back end of the engine apart," said Dale Eschelman, Stewart & Stevenson manager of the Metro parts and service account. "It was an eight- to ten-hour job per engine."

Dale found out about the engines at 4:30 p.m. on Friday. That evening, Stewart & Stevenson took them back to the shop where a crew worked through the weekend to get the engines ready for Monday morning. "We had to borrow parts from other engines and even fly some parts in from other Stewart & Stevenson locations," Dale said, "but we had those engines back at Metro by 7:30 a.m. on Monday. They didn't lose any time at all."

Dale spends part of nearly every day on-site at some Metro location and claims his success in getting their business, which normally ranges from $500,000 to $1 million a year, stems from being willing to "jump through the flaming hoop" whenever required. "We do the same thing for so many people every day of the week, it really isn't so special," he said.

But it did seem special to Russell Pentz, Metro's assistant general manager for maintenance, and the number two man in the organization which had recently been named the top large transit authority in the country by American Passenger Transportation Association. In the four years Pentz had been with Metro, he had watched the number of miles between road calls rise from 550 to more than 12,000 for the 1,084 buses Metro owned. And he credited Stewart & Stevenson with significant help for that improvement.

"Stewart & Stevenson is very aggressive in their help," Pentz said. ""They're big, yet they're small. They react to all my problems. If they can't answer it, they get someone who can. I consider them an extension of my own technical force. And Dale is super," he continue. "He's around at the drop of a hat, and he always manages to find the right people to talk to.

"I remember once we were in real bad trouble with transmissions. We weren't able to rebuild the product with the quality we needed in the numbers we needed. In 48 hours Stewart & Stevenson bailed us out with seven transmissions.

"They bend over backwards to provide training to our supervisors and mechanics. They've helped us develop a method to predict maintenance. They do a large majority of our warranty work. And in everything they do with us, there's a lot of give and take."

Pentz summed it up, "It's been a good marriage."

Dale agreed. He had just won a $2.5 million competitive bid to refurbish the power packages of 97 Metro coaches. Flaming hoops had their rewards.

Emerging A Global Business

While remaining loyal to its 50 year relationship with Detroit Diesel, and still the world's largest distributor of Diesel engines and parts, by 1988 Stewart & Stevenson had grown in new directions. The company developed and manufactured a power spectrum of original equipment products utilizing many makes of Diesel engines as prime movers. The company's products had become more complex and its markets more global.

Along with the use of a variety of Diesel engines from sources around the world, Stewart & Stevenson's sustained involvement with gas turbine technology was a major factor in the company's evolution into a global company that handled major power projects.

$106 Million Contract

In 1987 Stewart & Stevenson received a $106 million contract, the largest in the company's history, to supply power for the U.S. Air Force Peace-Shield project in Saudi Arabia. The generator sets would be assembled in the company's Saudi facility using Diesel engines supplied by a foreign manufacturer.

That same year, the company sold and delivered nine GE LM2500 gas turbine generators to Brazil for more than $76 million. The sale of an LM2500 to the Netherlands, scheduled for 1988 delivery, represented Stewart & Stevenson's first sale of a turbine generator in Europe. Ten Allison gas turbine generator sets were sold for the Iraqi pipeline.

The success of these and other similar sales was the catalyst for future growth as potential customers worldwide learned of Stewart & Stevenson's capabilities, which included packaged gas turbines for cogeneration, primary and backup power, and other applications. The company's power spectrum ranged from 500kW industrial gas turbines to 50,000kW aeroderivative units. The bulk of Stewart & Stevenson's gas turbine business centered around GE LM2500 and LM5000 turbines.

Stewart & Stevenson expanded its Jacintoport manufacturing facility, adding 28,600 square feet and upgrading full load test capability to include the larger LM6000 turbines.

This LM2500 gas turbine generator and related control system (a cogeneration package) undergoes a full load string test before shipment to a customer.

Stewart & Stevenson's advanced design blowout prevention control system.

With the acquisition of the assets of George Engine Co.. Stewart & Stevenson obtained a full-service branch in New Orleans, which added southern Louisiana, Mississippi and Alabama to its Detroit Diesel distribution territory.

More Expansion

At the end of 1987, Stewart & Stevenson was named the exclusive world-wide distributor for Garrett turbine engines, designed for mid-range applications. The company immediately established Garrett subdistributors and agents in Japan, Spain, Italy, Europe, Chile, Argentina, Venezuela and Saudi Arabia, and targeted India, Australia and Indonesia.

Nearly five decades after Stewart & Stevenson's original liaison with the Electro-Motive Division of General Motors Corporation ended, the company signed a distribution agreement with EMD for a ten-state area, as well as Mexico and Central America.

On January 12, 1988, Stewart & Stevenson announced its return to the Blowout Preventer Controls market. The company had been the acknowledged leader in this field until 1978 when the assets and businesses of the Oiltools Division were sold.

In 1988 the Distribution Division acquired the assets of George Engine Co., which gave Stewart & Stevenson a full service branch in New Orleans and added southern Louisiana, Mississippi and Alabama to the company's Detroit Diesel distribution territory. Located near the intercoastal waterway and Mississippi River, the facility included 1300 feet of dock space.

In December 1988, Stewart & Stevenson was appointed the distributor of Perkins Diesel engines for Texas, New Mexico and southern Louisiana. The Perkins line provided the company with a competitive engine for lower horsepower applications.

Also in 1988, Stewart & Stevenson formed the Accessory Product Sales Group to concentrate on accessory products that would complement the equip-

Bob H. O'Neal, President (1988-1990); President and CEO (1991-1997).

ment built or distributed by the company.

In 1989 Stewart & Stevenson was awarded the John Deere dealership for the entire upper Texas Gulf Coast, a 19-county area. This dealership included sales, service, parts and rentals for John Deere light industrial, construction and forestry equipment.

The John Deere dealership encompassed a 19-county area of the Upper Gulf Coast of Texas.

Stewart & Stevenson opened a branch in Brownsville, Texas, to serve the lower Gulf Coast in marine, light industrial, agricultural and commercial applications.

Bob H. O'Neal, President

Bob H. O'Neal joined Stewart & Stevenson in 1964 after a successful career as an Air Force officer. He worked his way through the ranks, serving as a salesman, branch operations manager, branch manager, and president of Stewart & Stevenson Power, Inc. in Denver before being elected President of Stewart & Stevenson Services, Inc. in 1988. He became President & CEO in 1991.

Under O'Neal's leadership, the Gas Turbine Products Division became the world leader in engineered power systems and the Distribution Division grew at an unprecedented rate, gaining more territories for Material Handling Construction Equipment and other products. In 1991, Stewart & Stevenson was named to the Fortune 500 list of Industrial Companies.

O'Neal retired from Stewart & Stevenson in 1997.

Revenues Reach and Exceed $500 Million

Revenues for 1988 climbed to $473.7 million; in 1989 to $604.9 million. Earnings in 1988 increased to $20.8 million; in 1989 to $27.3 million.

According to Stewart & Stevenson President Bob O'Neal, reporting in the March 1990 *S&S Today*, both divisions of the company — Engineered Power Systems and Distribution — contributed to the record sales and earnings. Growth leaders in 1989 were the Defense Products Group, with a whopping 235 percent increase in sales volume; California Cogen Operations & Maintenance Group with a 136 percent increase; the Gas Turbine Division with 42 percent increase; the Dallas Branch with 36 percent increase, and Houston Hyster Operation with 25 percent increase in sales volume.

The top five company operations with increased profitability over the prior year were California Cogen, with 190 percent; the New Orleans Branch, with 163 percent; Thermo King with 97 percent; the Beaumont Branch with 94 percent; and the Dallas Branch with a 71 percent increase.

Highlights of 1988 for Engineered Power Systems included shipment of the company's first LM5000 cogeneration package, introduction of a new de-icer to the company's line of aircraft ground support equipment, and delivery of $52 million of the $106 million Peace Shield contract for Diesel-powered generators.

During 1988, Stewart & Stevenson received orders for more than 500 megawatts of turbine capacity. The product line now concentrated on three major areas: generator sets in the 500kW to 5,000kW power range for applications including hospitals, hotels, office buildings and industrial facilities; large gas turbine packages from 20 to 50 megawatts utilized by utility companies

A 28 MW cogeneration plant.

These six Stewart & Stevenson 21 MW prime power generating plants were installed in South America.

A generator set is being installed in a high-rise condo in Houston.

to augment baseload capacity during peak usage periods; and large gas turbine systems for cogeneration plants that produced both electricity and process steam.

In 1989, Thomassen International and Stewart & Stevenson formed a joint venture to sell and service gas turbine-powered cogeneration and power-generating equipment in Europe and Africa. The alliance positioned Stewart & Stevenson as a strong contender in the European market for aeroderivative gas turbine generator sets.

By the end of the decade, gas turbine revenues accounted for two-thirds of all Engineered Power Systems' revenues and were approaching 50 percent of total company volume.

"It's a terribly efficient way of producing electricity," Chairman of the Board C. Jim Stewart II explained the gas turbine's success in a May 29, 1989 *Forbes* article. "You've got a tremendous amount of horsepower in a small package, and GE's working to get more—an LM 8000 would produce 100 megawatts, for example, though its dimensions would not be much larger than the 5000."

The LM 5000 gas turbine cogen packages, measuring 65 feet long by 14 feet wide by 30 feet high, were built, tested and assembled in the company's Jacintoport plant, then shipped on 12-axle trucks to the customer's site and put into operation in 30 to 60 days.

In 1990, General Electric introduced the new LM6000 gas turbine and selected Stewart & Stevenson as the exclusive packager of this engine within North America, Taiwan and Korea.

Chairman of the Board Stewart predicted in the *Forbes* article that gas turbines would soon dominate Stewart & Stevenson's business. But he noted that the company had other prospects as well.

New Military Contracts

A new $17 million military contract awarded the company in 1988 was for production of 15 prototype trucks for the Family of Medium Tactical Vehicles (FMTV), a series of 2.5 to 5-ton trucks. The prototypes were to be used in a three-company "fly-off" — a head-to-head test against competitors — for a contract that called for future production of 18,000 trucks with options for an additional 32,000 vehicles. The competition was to be conducted over a two to three-year period.

Stewart & Stevenson also engaged in a "fly-off" for a major Air Force contract for Light Electrical Generator Sets (LEGS) for air transport use. This competition, like the FMTV program, would result in a long term contract, for up to $100 million in equipment sales, for the winner. Prototype generators were delivered to the Air Force in 1989 for analysis.

Testing was underway in 1988 of Stewart & Stevenson's All Terrain Vehicle (ATV), a three or four-axle all-wheel drive truck designed to handle off-road conditions and maintain on-road travel at speeds up to 60 mph.

The company also developed a tow tractor for the U.S. Air Force and a 30-ton, hydraulic scissors loading unit for Navy and Air Force use.

Stewart & Stevenson's long-term relationship with the Navy continued with another contract for Allison gas turbines for use on Spruance-class destroyers. The company received a second contract in 1989 to build additional Diesel-powered generators for the Navy's Oliver Hazzard Perry-class Patrol Frigates, and also was building shipboard gas turbine generators for use aboard the Navy's new Aegis Cruisers.

The U-30 tow tractor for the Navy and Air Force.

Contracts for the Taiwanese Navy included a $6 million order for Diesel generators in 1989 and a $20 million order in 1990.

Stewart & Stevenson also was building 900 trailer mounted generators under the Army's Quiet Generator program, a contract that specified units capable of operating at 70 decibels at 7 meters. Extensive engineering was required to meet the stringent sound requirements. The generator sets would be used during the Gulf War.

Four 1000kW Diesel powered ship service generator sets were used aboard each Perry Class (FFG-7) patrol frigate.

This Quiet mobile generator set for the Army was part of a set of 5kW, 15kW and 30kW units.

The 40-foot Mercedes-built bus serves Houston Metro customers.

The Best Laid Plans...

Building on the company's history in the transportation industry that began in the 1920s, in 1987 Stewart & Stevenson began a new venture — the assembly and distribution of transit and shuttle buses, which research showed comprised a domestic market of approximately 5,000 units valued at $500 million annually.

In 1989, the Transit Product Division began marketing its 25-foot StarShip Shuttle and 30-foot Gemini buses, designed for airport and hotel shuttle service or urban transit, and the 40-foot Apollo series, designed for metropolitan transit use.

The buses were fabricated via a joint venture agreement with South American manufacturers such as Mercedes Benz and Marcopolo of Brazil. The bus bases were shipped to the U.S., where Stewart & Stevenson provided complete final assembly, including power trains, air conditioning, and interior furnishings depending upon the ultimate customer. For shuttle buses, which arrived basically complete, Stewart & Stevenson acted as distributor.

In keeping with the company's "single source responsibility," Stewart & Stevenson set up an extensive service network for the buses.

The product line provided an excellent example of the synergy Stewart & Stevenson sought between Engineered Power Systems, Distribution and its service and accessory business. The company's Detroit Diesel and Thermo King lines provided the power train and air-

Stewart & Stevenson's StarShip "Apollo" bus.

The 25-foot Starship shuttle bus built by Marcopolo.

Completely encased in silicon, Stewart & Stevenson's electronic gauge easily survived voltage spikes, ignored electro-magnetic interference and was easy to reprogram. The company's electronic systems for Automated Density Control and Automated Chemical Control on blenders sold for less than half the price of the competition's systems.

The 1992 acquisition of Quality Oilfield Products (QOP™) and Foley Valves™, a manufacturer of blowout preventers, high-pressure valves and other drilling industry products, enhanced Stewart & Stevenson's petroleum product line. Manufacturing capability acquired through QOP™ enabled Stewart & Stevenson to manufacture not only its new coiled-tubing injector head, but the entire truck- or trailer-mounted unit, including blowout preventers, hydraulics, controls and electronics. QOP™ also added ISO-9001 certified manufacturing capability in three new locations: Houston; Jennings, Louisiana and Arbroath, Scotland. By the end of the decade, Stewart & Stevenson had completed Phase I and Phase II of a new Petroleum manufacturing complex to house the Quality Oilfield Products and the Oilwell Service Equipment groups.

In 1994 Stewart & Stevenson delivered to a Norwegian drilling company a new snubbing unit rig that included the company's patent-pending blowout prevention stack with a shear ram designed, in the case of a blowout, to instantly cut off any downhole tools, close the well and hold the pressure. The $5 million rig carried out typical snubbing tasks, handled conventional workovers, performed coiled tubing work and drilling. The new blowout prevention technology was expanded to larger blowout preventers.

This meter, which can be programmed as a pressure, density, pH, viscosity or other current loop/process display, is used in the petroleum industry.

Stewart & Stevenson tests its proprietary sub-sea blowout preventer in the world's largest pressure test tank at the U.S. Navy's Subsurface Warfare Center in Annapolis, Maryland.

Global Marine Drilling Inc. used Stewart & Stevenson's proprietary marine riser system in 7,718 feet of water, setting a world record for ultra-deepwater drilling.

Stewart & Stevenson's first Rubicon rig, a combination drilling, workover and snubbing rig, worked flawlessly in the North Sea in 1996. The company designed a smaller, less costly version based on the same principles for jack-up workover rigs.

In 1997 the company introduced its proprietary marine riser system, developed over a two-year period, designed for deepwater drilling applications. Global Marine Drilling Inc. used the marine riser in 7,718 feet of water, setting a world's record for ultra-deepwater drilling.

The Petroleum Products Division in 1998 received three orders for well servicing equipment: two from Russian Petroleum Associations and one from China's Shengli Oilfield, for well fracturing equipment including high-pressure truck mounted pumpers and related support equipment, and for coiled tubing units, nitrogen units and related support equipment.

In 1998, Stewart & Stevenson acquired H&H Rubber Company, which manufactured proprietary rubber products to meet specifications of the Stewart & Stevenson line of blowout preventers.

To meet the growing demand, this new Petroleum Equipment Division plant was constructed in 1998.

The company also developed and tested a proprietary sub-sea blowout preventer (BOP) designed to withstand the extreme pressures of deepwater applications. An industry first, the BOP design was function-tested at 11,000 feet and found to be flawless in the largest pressure test tank in the world at the U.S. Navy's Subsurface Warfare Center in Annapolis, Maryland.

John "Tommy" Wall, who began his career at Stewart & Stevenson as a sales trainee in 1969, became vice president and general manager of Stewart & Stevenson's Petroleum Equipment Division. In 2000, the Petroleum Equipment Division approached $100 million in business annually.

Fifth Generation Stewarts Join the Company

C. Jim "Jamie" Stewart IV, son of C. Jim Stewart III and fifth generation Stewart, joined Stewart & Stevenson in August 1997 after graduating from Southern Methodist University. He began full-time work as a sales trainee in the Oilwell Services Equipment Department of the Petroleum Equipment Division.

His brother, Robert "Rob" Ross Stewart, joined the Oilfield Group as a sales trainee in June 1999 after his graduation from Texas Christian University.

Garth "Trip" C. Bates, son of Garth Bates and a fifth generation member of the Stewart family, joined the company in 1997 as a sales trainee in Denver, Colorado.

Meridith Mendenhall Powell, granddaughter of Tom Langham and a fifth generation Stewart family member, joined the company as an account representative for the Marketing Services Group.

Alternate Fuel Technologies Add Pizzazz to Distribution Division

Anticipating future tightening of the Clean Air Act, Stewart & Stevenson in 1988 had obtained funding from the Gas Research Institute to develop natural gas fueling for a standard Detroit Diesel engine. The work was carried out at Ortech, the Ontario Research Institute, located in Toronto. While there, the company met a group of engineers who were developing a natural gas system for automobiles and trucks. As a result of this work, in 1991 Stewart & Stevenson added two revolutionary new alternate fuel systems to the products the company distributed and serviced for Detroit Diesel Corp., Hyster, John Deere, Thermo King, Cooper-Superior, and Rail King®, the company's own proprietary product for in-plant rail car switching.

The two new alternate fuel systems developed in Toronto were the Detroit Diesel Pilot Ignition Natural Gas Engine and the Gaseous Fuel Injection (GFI) System.

Unlike gasoline engines, Diesel engines have no spark plug. Instead, they inject Diesel fuel into compressed air which spontaneously ignites, powering the cylinders. Diesel engines, particularly when not properly tuned, often emit visible soot. Stewart & Stevenson significantly improved that situation by adding natural gas to the Diesel engine. Diesel fuel still provided the ignition, hence the name "pilot ignition," but natural gas provided up to 85 percent of the total power. The result was increased efficiency, lower costs and much cleaner exhaust.

To accomplish this, Stewart & Stevenson used a standard Detroit Diesel engine, a modified gas valve, a solid-state valve controller and advanced electronics to tie it all together. The first prototype went into operation in 1990 for the Denver transit system. Hundreds of engines were sold with the conversion kit, which was offered as optional equipment by Detroit Diesel. If an engine ran out of compressed natural gas, it continued running on pure Diesel fuel.

The GFI system, developed by Ortech's electronic engineers, metered natural gas into virtually any automobile or light truck gasoline engine without modification of the engine. The closed loop, solid state system was 50 percent smaller than any existing mechanical system, with 65 percent fewer components. Natural gas, which burned cleaner than gasoline, saved 30 to 60 percent in fuel costs. Unlike the natural gas system developed by automakers, which ran only on natural gas, the Stewart & Stevenson GFI system would use either natural gas or gasoline. A simple switch on the dash allowed the driver to switch between fuels as needed. In 1992, the company set up nationwide distribution for the GFI System by forming a jointly owned subsidiary, GFI Control Systems, with Devtek Corporation, a Canadian company. Stewart & Stevenson sold its interest in GFI to its partner in 1999.

Other Manufactured Products

Other products developed and manufactured by Stewart & Stevenson included Rail King® rail car movers, Powerlift™ wheelchair lifts and RSS snowblowers, which were capable of hurling snow up to 100 feet at a rate of 5,000 tons per hour, making them ideal for airports and highways. In 1993 a new line of snow "brooms" manufactured by Stewart & Stevenson Power to designs from Schorlling in Hannover, Germany were added. Several machines operating in tandem could clean an airport runway of snow or other debris in minutes.

Introduced in the early 1990s were the Detroit Diesel Series 60 engine, which captured a sizable share of the Class 8 truck market; Allison's World Transmission, and EMD's new turbo-charged 710 Series marine engines, which produced up to 4,500 horsepower. As the North American Free Trade Agreement took effect, Stewart & Stevenson's distribution territory became the hub of trade with Mexico, since it was located on the major trucking corridor between the two countries.

Snowblowers developed by Stewart & Stevenson (Denver) can hurl snow 100 feet at a rate of 5,000 tons per hour. They can clean an airport runway of heavy snow and ice or other debris in minutes.

The Rail King® rail car mover is part of the TUG® Ground Support Division product line.

The Waukesha 12VAT25 Lean Burn gas compression package.

In 1993, the Central Parts Distribution Center was closed and replaced by three regional parts warehouses for Detroit Diesel, EMD and accessory parts line. They were located in New Orleans, Dallas and Houston, the company's three largest branch operations.

Additional distributorship territory in South America awarded to Stewart & Stevenson for EMD engines included Venezuela, Guyana, Surinane, French Guiana, Peru, Bolivia, Paraguay, and Argentina.

Added to the company's full line of John Deere construction equipment was a new line of composting equipment, built by Diamond Z, Banner Manufacturing and Scarab Manufacturing. It turned tree stumps and other "green" wastes into compostable material.

At the end of 1994, Stewart & Stevenson acquired the assets of PAMCO®, a $40-million company specializing in the distribution and packaging of Waukesha engine products for the petroleum and industrial marketplace. PAMCO®, which became a part of Stewart & Stevenson Power, Inc., added natural gas-fired engines to the company's product line, along with a distribution territory that covered virtually the entire western half of the U.S. It included an alliance with Waukesha Engine Products Division of Dresser Industries, Inc. PAMCO® added proven technology in both gas compression and power generation.

A new relationship between Detroit Diesel Corp. and the master distributor for MTU (Motoren und Turbinen-Union Friedrichshafen gmbH) in the United States made available to Stewart & Stevenson a line of high-tech, heavy-duty engines.

Two new types of equipment added to Distribution's

Robert Hargrave, President & CEO (1997-1999).

Construction Equipment division in 1994 were shoring systems for underground construction and skid-steer loaders. A new 50,000 square-foot facility on 22 acres in east Houston provided room for an extensive lease inventory. It included 32 service bays and the advanced technology required to provide the best possible customer service.

The Material Handling Division added a new narrow-aisle forklift unit to its product line and a chemical repair system for warehouse floors.

Standards of Business Practices

Employees of Stewart & Stevenson since the company's beginning have taken pride in the company's ethical standards and operations. In the early 1950s, management had drawn up a list of Basic Principles, which guided the company and its employees over the years. In 1995, Stewart & Stevenson formalized the company's ethical beliefs and standards further by updating the early document and publishing a "Standards of Business Practices" handbook. Comprehensive ethics training was provided to every employee and a Business Practices office, directed by corporate vice president T. Michael Andrews, was established to provide guidance on making sound business decisions. This service continues to be made available to all employees, suppliers and customers through a worldwide helpline.

The Stewart & Stevenson Business Practices program has become a model used by the Air Force for other vendors to follow.

Robert L. Hargrave, President

Robert L. "Bob" Hargrave, Stewart & Stevenson's Chief Financial Officer, was appointed acting Chief Executive Officer in 1997 and President in 1998. Hargrave had been with the company for 28 years, serving as manager of general accounting, assistant controller, treasurer and chief financial officer.

In January 1999, after 31 years of dedicated service, Robert L. Hargrave resigned as President, Chief Executive Officer and member of the Board of Directors. Chairman of the Board C. Jim Stewart II once again became Acting CEO until the search for a new president was concluded. Organizational changes followed.

Sales, Acquisitions, Reorganization Ready Company for New Century

Gas Turbine Division Sold

In January 1998 Stewart & Stevenson sold the Gas Turbine Division to General Electric for $600 million in cash. Major changes in the electrical power generation market worldwide led to the sale. The market, which utilized gas turbines, was disrupted domestically by deregulation and internationally by privatization, factors which depressed demand and thus the selling price of turbines. During this transitional period, consolidation in the power generation sector added to the competitiveness of the market, and Stewart & Stevenson opted to become a seller in the consolidation trend.

Funds from the sale permitted Stewart & Stevenson to pursue new opportunities in Diesel and gas-fired reciprocating engine markets through strategic acquisitions. The company also reduced bank debt by $260 million and initiated a $120 million open market stock repurchase program.

Distribution Renamed Power Products

At the suggestion of the Marketing Department, in 1996 Stewart & Stevenson's Distribution segment was renamed Power Products to reflect its core business. It set an aggressive goal for $750 million in sales by the year 2000, a target to be achieved through a combination of internal growth and acquisitions.

In April 1997 the company acquired Sierra Detroit Diesel, a distributor for Detroit Diesel, Waukesha and Allison products in northern California. It added more than $30 million to 1997 sales and gave Stewart & Stevenson two new locations on the West Coast, San Leandro and Sacramento, California. Stewart & Stevenson in 1997 also was appointed distributor of Detroit Diesel and Waukesha engines for Colombia, South America.

Because it did not fit with the company's long-range strategic goals, the John Deere construction equipment dealership that covered 19 counties in and around Houston was sold in 1997 for $30 million cash. Stewart & Stevenson's remaining John Deere dealership in Wyoming operated as a product line within the facilities shared by the Detroit Diesel and Waukesha distributorships.

In 1998, Stewart & Stevenson acquired IPSCO Co., Inc., the Deutz distributor for Louisiana, Mississippi, Arkansas and Western Tennessee. The company also acquired the Deutz distributorship franchise for Texas, Oklahoma and Kansas from Harley Equipment Company.

In December 1999, Stewart & Stevenson acquired the assets of Thermo King of Northern California, a Thermo King dealership with locations in Sacramento, San Leandro and Oakland, California. The dealership, which had 1999 sales of approximately $12 million, was to be operated by Stewart & Stevenson's wholly owned subsidiary, Sierra Detroit Diesel Allison, Inc., which also had facilities in Sacramento and San Leandro. For 29 years, Stewart & Stevenson had been a Thermo King dealer, with dealerships located in Texas and Louisiana. Thermo King is the premier manufacturer of refrigeration equipment utilized in trucks, trailers, buses, marine containers and other applications.

PAMCO Services International

Stewart & Stevenson acquired PAMCO®, the western U.S. engine distributor for Waukesha gas compression and generating equipment, and Compression Specialties, Inc. (CSI), a small gas compression rental and service company in Wyoming, which strengthened the company's presence in the Rocky Mountain region. In March 1998 CSI was renamed PAMCO Services International (PSI®). This new subsidiary was formed to sell, rent and lease gas compression equipment, and to provide complementary maintenance and operational services.

PSI® was formed to sell and lease gas compression equipment and to provide complementary maintenance and operational services.

Michael Grimes, President & CEO (1999 - Present)

Business Segments Reorganized

In response to the sale of the Gas Turbine Operations and the requirements of Financial Accounting Standards Board Statement No. 131, Stewart & Stevenson reorganized its business segments in 1998 to include Power Products, Tactical Vehicle Systems and Petroleum Equipment. Business activities not defined as a specific segment included aircraft ground support equipment, leased gas compression equipment and related services, and other miscellaneous sales.

Michael L. Grimes, President

Michael L. Grimes was hired as President and CEO in April 1999. He came to Stewart & Stevenson from Cooper Cameron Corporation where he was President of Cooper Cameron Power Generation, and had served as President of Cooper Energy Services from 1996 to 1998. He previously had been employed by General Electric Company where he spent 23 years in various managerial positions, primarily in GE's power generation and service business.

New Leadership for New Era

"With this new leadership, we are confident that Stewart & Stevenson is headed into a new era, well prepared to meet the challenges of the next century," Board Chairman C. Jim Stewart II proclaims. "This change in management brings a new management style that will make Stewart & Stevenson a more sophisticated publicly held company than we have been in the past. Yet we expect to retain the important aspects of our tradition as a family business: The Can Do Attitude, The Customer Comes First, and Giving Employees The Support They Need."

As we go to press in January 2001 with this 100th anniversary volume, Stewart & Stevenson, now a billion-dollar, global corporation, and its more than 4,200 employees look toward its second century with confidence and pride.

Epilogue: The Next Century

As I am writing these closing comments, the one certainty that I have for the next 100 years is a continued rate of accelerated change. We are just beginning to explore the new cyber frontier. Communications technologies and computer networks — particularly the Internet — are creating an emerging vast new market place and changing fundamental concepts of space, time and organization. Although the full consequences of this new cyber technology cannot now be fully imagined, shared knowledge through communications technology will be a dominant force in 21st century economics and commerce. Concepts such as E-business, virtual organizations, supply chain productivity, virtual inventory, and a host of others that have yet to take their final form will profoundly reshape our world business environment.

One hundred years from now when the Stewart & Stevenson CEO in the year 2100 is preparing a similar article, she or he will know the answers to the many questions that we have today — just as we now have the answers to many of the questions that C. Jim Stewart and Joe R. Stevenson had when they founded this great company.

While we wrestle with the challenges and opportunities of our changing business environment, just as our founders did with their environment in the 1900s, there are a number of business elements that remain the same forever — past, present and future. The most important of these elements is the reality that everything in business begins and ends with a *Customer.* Whether customers are global, regional, or local, and products offered are equipment, service, or integrated systems, our company will remain focused on a true commitment to helping our customers meet their objectives through innovative solutions to their needs.

The cornerstone of this innovation comes from well-trained and motivated employees, our *Shared Values*, and our many successful supply partnerships. These elements have served us well and will continue to serve us successfully as we put our customers first in everything we do, while serving the best interests of our shareholders.

This is an exciting time for Stewart & Stevenson. I am very fortunate to have been selected to provide CEO leadership as we head into this new century. The next 100 years promises to be a challenging opportunity where we will continue our tradition of meeting our customers' needs. I am confident in this great company's ability to meet these challenges of the future as it has for the last 100 years.

Michael L. Grimes

The Power of Generations.

THE FAMILY SCRAPBOOK. An archive of memories and people that lives on forever. It's where grandchildren learn about the heritage of generations past, and see their heroes come to life before their very eyes.

Stewart & Stevenson approaches our 100th year of providing engineered power to the world by reflecting on the generations of our employees who have dedicated their services to generations of our customers. As we continue to build our company on the strengths of diversification, we will never overlook The Power of Generations.

ENGINEERED POWER STEWART & STEVENSON

WORLD HEADQUARTERS
P.O. Box 1637/Houston, Texas 77251-1637 (713)868-7700

The History of the Stewart & Stevenson Logo

One of the industry's most recognized corporate symbols, the Stewart & Stevenson logo has been synonymous with quality products for almost 100 years.

The first Stewart & Stevenson trademark was registered in 1905. An official trademark was the means by which the company could provide positive protection against copying and deception.

C Jim Stewart & Stevenson
Est. 1902
Houston, Texas

The first trademark shows the words "C. Jim Stewart & Stevenson" with a horseshoe-shaped "C," scrolled initial letters and a butterfly ampersand. The lettering, typical of the era, reflected the promise of the craftsmanship owners C. Jim Stewart and Joe R. Stevenson offered to their horse and carriage customers, as well as pride in their business partnership.

As the "horseless carriage" became more prominent, and the company's craftsmanship was extended to automobile repair, parts manufacturing and a Marion-Handley dealership, the company's logo was changed to reflect its expanded business. An auto was inserted into the "C" of the logo. The image depicted was later updated to reflect the company's addition of truck-body building to its trade.

C Jim Stewart & Stevenson
HEAVY AUTO ACCESSORIES
TRIMMING, PAINTING, BODY BUILDING
PHONE PRES. 0434
HOUSTON, TEXAS

Upon adding its first operating division, the Alemite Company of the Gulf Coast in 1923, the C. Jim Stewart & Stevenson logo acquired an altogether new look. An upward-aimed "Stewart & Stevenson" banner was centered over an all-inclusive oval that could separately depict the diversified divisions of the company that rapidly came into being: Distributing, Automotive Supply, and Hertz Truck Lease. The block-style lettering represented a company built upon a firm and lasting foundation.

ALEMITE CO. OF THE GULF COAST

STEWART & STEVENSON DISTRIBUTING CO.

C Jim STEWART & STEVENSON

The acquisition of the General Motors Detroit Diesel franchise in 1938 resulted in the formation of the Engine Division, and the logo's oval soon depicted the engines and engine applications Stewart & Stevenson developed for the agriculture, boating, and oilfield industry, as well as the day and night parts and service availability the company afforded every customer.

The formation of Stewart & Stevenson Services to fulfill government contracts for the massive war effort during World War II added engines to the logo's oval.

STEWART & STEVENSON SERVICES

After World War II ended, the original partnership became four companies: C. Jim Stewart & Stevenson, Inc. (the original C. Jim Stewart & Stevenson), Stewart & Stevenson Services, Inc. (formerly the Engine Division), Stewart & Stevenson Distributing Company, and Alemite Company of the Gulf Coast.

C. Jim Stewart & Stevenson, Inc. expanded into new arenas after the war. Paul Koomey's development of a new blowout preventer closing unit led to the formation of the Oilfield Equipment Division in 1954, The Material Handling Division represented Hyster and Raymond forklifts. The Truck Lease Division grew to be the largest in the Southwest. Machinery Acceptance Corporation was created to finance purchases of engine applications.

Stewart & Stevenson Services, which represented 80 percent of the combined companies' business, expanded rapidly, adding franchises for other engines and branches for sales and service across Texas and beyond. "Finding a better way" became the company's new marching order, and research and engineering created numerous applications for the engines the company represented. "Engineered Power," applied to the oil field, airline ground support, missile and space program, marine industry, agriculture industry, military and for emergency backup applications everywhere, was backed up with the company's "Parts & Service...Anytime — Anywhere"® promise. By 1960, the company's engine applications were being utilized internationally, as the logo depicted on the annual report that year indicated.

In 1972, the allied companies of Stewart & Stevenson were reorganized, with Stewart & Stevenson Services, Inc. becoming the parent company. By this point in the company's history, its products were being utilized around the world, and a new logo was developed to represent that fact. Inside the oval appeared the Western hemisphere of the globe, with a portion of the Eastern hemisphere appearing on the eastern horizon. The banner with bold Stewart & Stevenson lettering, still aimed upward, now had a more pointed end. The new logo graced the cover of the 1974 annual report. In 1976 Stewart & Stevenson went public.

Continuing to build on its proven foundation of quality products, goodwill, and honest and intelligent service, Stewart & Stevenson further diversified and rapidly expanded its operations worldwide. The grid of the globe was added to the logo to symbolize its new status as a global operation as the company moved into world headquarters at 2707 North Loop West in Houston.

At the dawn of the new millennium, Stewart & Stevenson remains a global organization dedicated to providing products and services that reflect the world's ever-changing technology and meet the needs of its worldwide customers. The Stewart & Stevenson logo is a bold symbol of the company — one that is proud of its past, preserving and building upon its heritage, but simultaneously looking toward the future for new opportunities. Under this banner, Stewart & Stevenson moves forward into its second century with power and vision.

INDEX